Introduction to Excel®

David C. Kuncicky

Department of Electrical Engineering
FAMU-FSU College of Engineering

Prentice Hall
Upper Saddle River, NJ 07458

Library of Congress Information Available

Editor-in-chief: **MARCIA HORTON**
Acquisitions editor: **ERIC SVENDSEN**
Director of production and manufacturing: **DAVID W. RICCARDI**
Managing editor: **EILEEN CLARK**
Editorial/production supervision: **ROSE KERNAN**
Cover director: **JAYNE CONTE**
Creative director: **AMY ROSEN**
Manufacturing buyer: **PAT BROWN**
Editorial assistant: **GRIFFIN CABLE**

©1999 by Prentice-Hall, Inc.
Simon & Schuster / A Viacom Company
Upper Saddle River, New Jersey 07458

Printed in the United States of America

10 9 8 7 6 5 4 3 2 1

This text was written and designed to
be used with Excel 97. Microsoft Excel
97® and Excel® are registered trade-
marks of the Microsoft Corporation.

ISBN 013-254749-X

PRENTICE-HALL INTERNATIONAL (UK) LIMITED, *London*
PRENTICE-HALL OF AUSTRALIA PTY. LIMITED, *Sydney*
PRENTICE-HALL CANADA INC., *Toronto*
PRENTICE-HALL HISPANOAMERICANA, S.A., *Mexico*
PRENTICE-HALL OF INDIA PRIVATE LIMITED, *New Delhi*
PRENTICE-HALL OF JAPAN, INC., *Tokyo*
SIMON & SCHUSTER ASIA PTE. LTD., *Singapore*
EDITORA PRENTICE-HALL DO BRASIL, LTDA., *Rio de Janeiro*

About ESource

The Challenge

Professors who teach the Introductory/First-Year Engineering course popular at most engineering schools have a unique challenge—teaching a course defined by a changing curriculum. The first-year engineering course is different from any other engineering course in that there is no real cannon that defines the course content. It is not like Engineering Mechanics or Circuit Theory where a consistent set of topics define the course. Instead, the introductory engineering course is most often defined by the creativity of professors and students, and the specific needs of a college or university each semester. Faculty involved in this course typically put extra effort into it, and it shows in the uniqueness of each course at each school.

Choosing a textbook can be a challenge for unique courses. Most freshmen require some sort of reference material to help them through their first semesters as a college student. But because faculty put such a strong mark on their course, they often have a difficult time finding the right mix of materials for their course and often have to go without a text, or with one that does not really fit. Conventional textbooks are far too static for the typical specialization of the first-year course. How do you find the perfect text for your course that will support your students educational needs, but give you the flexibility to maximize the potential of your course?

ESource—The Prentice Hall Engineering Source
http://emissary.prenhall.com/esource

Prentice Hall created ESource—The Prentice-Hall Engineering Source—to give professors the power to harness the full potential of their text and their freshman/first year engineering course. In today's technologically advanced world, why settle for a book that isn't perfect for your course? Why not have a book that has the exact blend of topics that you want to cover with your students?

More then just a collection of books, ESource is a unique publishing system revolving around the ESource website—http://emissary.prenhall.com/esource/. ESource enables you to put your stamp on your book just as you do your course. It lets you:

Control   You choose exactly what chapters or sections are in your book and in what order they appear. Of course, you can choose the entire book if you'd like and stay with the authors original order.

Optimize   Get the most from your book and your course. ESource lets you produce the optimal text for your students needs.

Customize   You can add your own material anywhere in your text's presentation, and your final product will arrive at your bookstore as a professionally formatted text.

ESource Content

All the content in ESource was written by educators specifically for freshman/first-year students. Authors tried to strike a balanced level of presentation, one that was not either too formulaic and trivial, but not focusing heavily on advanced topics that most introductory students will not encounter until later classes. A developmental editor reviewed the books and made sure that every text was written at the appropriate level, and that the books featured a balanced presentation. Because many professors do not have extensive time to cover these topics in the classroom, authors prepared each text with the idea that many students would use it for self-instruction and independent study. Students should be able to use this content to learn the software tool or subject on their own.

While authors had the freedom to write texts in a style appropriate to their particular subject, all followed certain guidelines created to promote the consistency a text needs. Namely, every chapter opens with a clear set of objectives to lead students into the chapter. Each chapter also contains practice problems that tests a student's skill at performing the tasks they have just learned. Chapters close with extra practice questions and a list of key terms for reference. Authors tried to focus on motivating applications that demonstrate how engineers work in the real world, and included these applications throughout the text in various chapter openers, examples, and problem material. Specific Engineering and Science **Application Boxes** are also located throughout the texts, and focus on a specific application and demonstrating its solution.

Because students often have an adjustment from high school to college, each book contains several **Professional Success Boxes** specifically designed to provide advice on college study skills. Each author has worked to provide students with tips and techniques that help a student better understand the material, and avoid common pitfalls or problems first-year students often have. In addition, this series contains an entire book titled *Engineering Success* by Peter Schiavone of the University of Alberta intended to expose students quickly to what it takes to be an engineering student.

Creating Your Book

Using ESource is simple. You preview the content either on-line or through examination copies of the books you can request on-line, from your PH sales rep, or by calling(1-800-526-0485). Create an on-line outline of the content you want in the order you want using ESource's simple interface. Either type or cut and paste your own material and insert it into the text flow. You can preview the overall organization of the text you've created at anytime (please note, since this preview is immediate, it comes unformatted.), then press another button and receive an order number for your own custom book . If you are not ready to order, do nothing—ESource will save your work. You can come back at any time and change, re-arrange, or add more material to your creation. You are in control. Once you're finished and you have an ISBN, give it to your bookstore and your book will arrive on their shelves six weeks after the order. Your custom desk copies with their instructor supplements will arrive at your address at the same time.

To learn more about this new system for creating the perfect textbook, go to **http://emissary.prenhall.com/esource/**. You can either go through the on-line walkthrough of how to create a book, or experiment yourself.

Community

ESource has two other areas designed to promote the exchange of information among the introductory engineering community, the Faculty and the Student Centers. Created and maintained with the help of Dale Calkins, an Associate Professor at the University of Washington, these areas contain a wealth of useful information and tools. You can preview outlines created by other schools and can see how others organize their courses. Read a monthly article discussing important topics in the curriculum. You can post your own material and share it with others, as well as use what others have posted in your own documents. Communicate with our authors about their books and make suggestions for improvement. Comment about your course and ask for information from others professors. Create an on-line syllabus using our custom syllabus builder. Browse Prentice Hall's catalog and order titles from your sales rep. Tell us new features that we need to add to the site to make it more useful.

Supplements

Adopters of ESource receive an instructor's CD that includes solutions as well as professor and student code for all the books in the series. This CD also contains approximately **350 Powerpoint Transparencies** created by Jack Leifer—of University South Carolina—Aiken. Professors can either follow these transparencies as pre-prepared lectures or use them as the basis for their own custom presentations. In addition, look to the web site to find materials from other schools that you can download and use in your own course.

Titles in the ESource Series

Introduction to Unix
0-13-095135-8
David L. Schwartz

Introduction to Maple
0-13-095-133-1
David L. Schwartz

Introduction to Word
0-13-254764-3
David C. Kuncicky

Introduction to Excel
0-13-254749-X
David C. Kuncicky

Introduction to MathCAD
0-13-937493-0
Ronald W. Larsen

Introduction to AutoCAD, R. 14
0-13-011001-9
Mark Dix and Paul Riley

Introduction to the Internet, 2/e
0-13-011037-X
Scott D. James

Design Concepts for Engineers
0-13-081369-9
Mark N. Horenstein

Engineering Design—A Day in the Life of Four Engineers
0-13-660242–8
Mark N. Horenstein

Engineering Ethics
0-13-784224-4
Charles B. Fleddermann

Engineering Success
0-13-080859-8
Peter Schiavone

Mathematics Review
0-13-011501-0
Peter Schiavone

Introduction to ANSI C
0-13-011854-0
Dolores Etter

Introduction to C++
0-13-011855-9
Dolores Etter

Introduction to MATLAB
0-13-013149-0
Dolores Etter

Introduction to FORTRAN 90
0-13-013146-6
Larry Nyhoff & Sanford Leestma

About the Authors

No project could ever come to pass without a group of authors who have the vision and the courage to turn a stack of blank paper into a book. The authors in this series worked diligently to produce their books, provide the building blocks of the series.

Delores M. Etter is a Professor of Electrical and Computer Engineering at the University of Colorado. Dr. Etter was a faculty member at the University of New Mexico and also a Visiting Professor at Stanford University. Dr. Etter was responsible for the Freshman Engineering Program at the University of New Mexico and is active in the Integrated Teaching Laboratory at the University of Colorado. She was elected a Fellow of the Institute of Electrical and Electronic Engineers for her contributions to education and for her technical leadership in digital signal processing. IN addition to writing best-selling textbooks for engineering computing, Dr. Etter has also published research in the area of adaptive signal processing.

Sanford Leestma is a Professor of Mathematics and Computer Science at Calvin College, and received his Ph.D from New Mexico State University. He has been the long time co-author of successful textbooks on Fortran, Pascal, and data structures in Pascal. His current research interests are in the areas of algorithms and numerical compuitation.

Larry Nyhoff is a Professor of Mathematics and Computer Science at Calvin College. After doing bachelors work at Calvin, and Masters work at Michigan, he received a Ph.D. from Michigan State and also did graduate work in computer science at Western Michigan. Dr. Nyhoff has taught at Calvin for the past 34 years—mathematics at first and computer science for the past several years. He has co-authored several computer science textbooks

since 1981 including titles on Fortran and C++, as well as a brand new title on Data Structures in C++.

Acknowledgments: We express our sincere appreciation to all who helped in the preparation of this module, especially our acquisitions editor Alan Apt, managing editor Laura Steele, development editor Sandra Chavez, and production editor Judy Winthrop. We also thank Larry Genalo for several examples and exercises and Erin Fulp for the Internet address application in Chapter 10. We apprcciate the insightful review provided by Bart Childs. We thank our families—Shar, Jeff, Dawn, Rebecca, Megan, Sara, Greg, Julie, Joshua, Derek, Tom, Joan; Marge, Michelle, Sandy, Lori, Michael—for being patient and understanding. We thank God for allowing us to write this text.

Mark Dix began working with AutoCAD in 1985 as a programmer for CAD Support Associates, Inc. He helped design a system for creating estimates and bills of material directly from AutoCAD drawing databases for use in the automated conveyor industry. This system became the basis for systems still widely in use today. In 1986 he began collaborating with Paul Rilcy to create AutoCAD training materials, combining Riley's background in industrial design and training with Dix' s background in writing, curriculum development, and programming. Dix and Riley have created tutorial and teaching methods for every AutoCAD release since Version 2.5. Mr. Dix has a Master of Arts in Teaching from Cornell University and a Masters of Education from the University of Massachusetts. He is currently the Director of Dearborn Academy High School in Arlington, Massachusetts.

Paul Riley is an author, instructor, and designer specializing in graphics and design for multimedia. He is a founding partner of CAD Support Associates, a contract service and professional training organization for computer-aided design. His 15 years of business experience and 20 years of teaching experience are supported by degrees in education and computer science. Paul has taught AutoCAD at the University of Massachusetts at Lowell and is presently teaching AutoCAD at Mt. Ida College in Newton, Massachusetts. He has developed a program, Computer-Aided Design for Professionals that is highly regarded by corporate clients and has been an ongoing success since 1982.

David I. Schwartz is a Lecturer at SUNY-Buffalo who teaches freshman and first-year engineering, and has a Ph.D from SUNY-Buffalo in Civil Engineering. Schwartz originally became interested in Civil engineering out of an interest in building grand structures, but has also pursued other academic interests including artificial intelligence and applied mathematics. He became interested in Unix and Maple through their application to his research, and eventually jumped at the chance to teach these subjects to students. He tries to teach his students to become incremental learners and encourages frequent practice to master a subject, and gain the maturity and confidence to tackle other subjects independently. In his spare time, Schwartz is an avid musician and plays drums in a variety of bands.

Acknowledgments: I would like to thank the entire School of Engineering and Applied Science at the State University of New York at Buffalo for the opportunity to teach not only my students, but myself as well; all my EAS140 students, without whom this book would not be possible—thanks for slugging through my lab packets; Andrea Au, Eric Svendsen, and Elizabeth Wood at Prentice Hall for advising and encouraging me as well as wading through my blizzard of e-mail; Linda and Tony for starting the whole thing in the first place; Rogil Camama, Linda Chattin, Stuart Chen, Jeffrey Chottiner, Roger Christian, Anthony Dalessio, Eugene DeMaitre, Dawn Halvorsen, Thomas Hill, Michael Lamanna, Nate "X" Patwardhan, Durvejai Sheobaran, "Able" Alan Somlo, Ben Stein, Craig Sutton, Barbara Umiker, and Chester "JC" Zeshonski for making this book a reality; Ewa Arrasjid, "Corky" Brunskill, Bob Meyer, and Dave Yearke at "the Department Formerly Known as ECS" for all their friendship, advice, and respect; Jeff, Tony, Forrest, and Mike for the interviews; and, Michael Ryan and Warren Thomas for believing in me.

Ronald W. Larsen is an Associate Professor in Chemical Engineering at Montana State University, and received his Ph.D from the Pennsylvania State University. Larsen was initially attracted to engineering because he felt it was a serving profession, and because engineers are often called on to eliminate dull and routine tasks. He also enjoys the fact that engineering rewards creativity and presents constant challenges. Larsen feels that teaching large sections of students is one of the most challenging tasks he has ever encountered because it enhances the importance of effective communication. He has drawn on a two year experince teaching courses in Mongolia through an interpreter to improve his skills in the classroom. Larsen sees software as one of the changes that has the potential to radically alter the way engineers work, and his book Introduction to Mathcad was written to help young engineers prepare to be productive in an ever-changing workplace.

Acknowledgments: To my students at Montana State University who have endured the rough drafts and typos, and who still allow me to experiment with their classes—my sincere thanks.

Peter Schiavone is a professor and student advisor in the Department of Mechanical Engineering at the University of Alberta. He received his Ph.D. from the University of Strathclyde, U.K. in 1988. He has authored several books in the area of study skills and academic success as well as numerous papers in scientific research journals.

Before starting his career in academia, Dr. Schiavone worked in the private sector for Smith's Industries (Aerospace and Defence Systems Company) and Marconi Instruments in several different areas of engineering including aerospace, systems and software engineering. During that time he developed an interest in engineering research and the applications of mathematics and the physical sciences to solving real-world engineering problems.

His love for teaching brought him to the academic world. He founded the first Mathematics Resource Center at the University of Alberta: a unit designed specifically to teach high school students the necessary survival skills in mathematics and the physical sciences required for first-year engineering. This led to the Students' Union Gold Key award for outstanding contributions to the University and to the community at large.

Dr. Schiavone lectures regularly to freshman engineering students, high school teachers, and new professors on all aspects of engineering success, in particular, maximizing students' academic performance. He wrote the book *Engineering Success* in order to share with you the *secrets of success in engineering study*: the most effective, tried and tested methods used by the most successful engineering students.

Acknowledgments: I'd like to acknowledge the contributions of: Eric Svendsen, for his encouragement and support; Richard Felder for being such an inspiration; the many students who shared their experiences of first-year engineering—both good and bad; and finally, my wife Linda for her continued support and for giving me Conan.

Scott D. James is a staff lecturer at Kettering University (formerly GMI Engineering & Management Institute) in Flint, Michigan. He is currently pursuing a Ph.D. in Systems Engineering with an emphasis on software engineering and computer-integrated manufacturing. Scott decided on writing textbooks after he found a void in the books that were available. "I really wanted a book that showed how to do things in good detail but in a clear and concise way. Many of the books on the market are full of fluff and force you to dig out the really important facts." Scott decided on teaching as a profession after several years in the computer industry. "I thought that it was really important to know what it was like outside of academia. I wanted to provide students with classes that were up to date and provide the information that is really used and needed."

Acknowledgments: Scott would like to acknowledge his family for the time to work on the text and his students and peers at Kettering who offered helpful critique of the materials that eventually became the book.

David C. Kuncicky is a native Floridian. He earned his Baccalaureate in psychology, Master's in computer science, and Ph.D. in computer science from Florida State University. He is also the author of *Excel 97 for Engineers*. Dr. Kuncicky is the Director of Computing and Multimedia Services for the FAMU-FSU College of Engineering. He also serves as a faculty member in the Department of Electrical Engineering. He has taught computer science and computer engineering courses for the past 15 years. He has published research in the areas of intelligent hybrid systems and neural networks. He is actively involved in the education of computer and network system administrators and is a leader in the area of technology-based curriculum delivery.

Acknowledgments: Thanks to Steffie and Helen for putting up with my late nights and long weekends at the computer. Thanks also to the helpful and insightful technical reviews by the following people: Jerry Ralya, Kathy Kitto of Western Washington University, Avi Singhal of Arizona State University, and Thomas Hill of the State University of New York at Buffalo. I appreciate the patience of Eric Svendsen and Rose Kernan of Prentice Hall for gently guiding me through this project. Finally, thanks to Dean C.J. Chen for providing continued tutelage and support.

Mark Horenstein is an Associate Professor in the Electrical and Computer Engineering Department at Boston University. He received his Bachelors in Electrical Engineering in 1973 from Massachusetts Institute of Technology, his Masters in Electrical Engineering in 1975

from University of California at Berkeley, and his Ph.D. in Electrical Engineering in 1978 from Massachusetts Institute of Technology. Professor Horenstein's research interests are in applied electrostatics and electromagnetics as well as microelectronics, including sensors, instrumentation, and measurement. His research deals with the simulation, test, and measurement of electromagnetic fields. Some topics include electrostatics in manufacturing processes, electrostatic instrumentation, EOS/ESD control, and electromagnetic wave propagation.

Professor Horenstein designed and developed a class at Boston University, which he now teaches entitled Senior Design Project (ENG SC 466). In this course, the student gets real engineering design experience by working for a virtual company, created by Professor Horenstein, that does real projects for outside companies—almost like an apprenticeship. Once in "the company" (Xebec Technologies), the student is assigned to an engineering team of 3-4 persons. A series of potential customers are recruited, from which the team must accept an engineering project. The team must develop a working prototype deliverable engineering system that serves the need of the customer. More than one team may be assigned to the same project, in which case there is competition for the customer's business.

Acknowledgements: Several individuals contributed to the ideas and concepts presented in Design Principles for Engineers. The concept of the Peak Performance design competition, which forms a cornerstone of the book, originated with Professor James Bethune of Boston University. Professor Bethune has been instrumental in conceiving of and running Peak Performance each year and has been the inspiration behind many of the design concepts associated with it. He also provided helpful information on dimensions and tolerance. Several of the ideas presented in the book, particularly the topics on brainstorming and teamwork, were gleaned from a workshop on engineering design help bi-annually by Professor Charles Lovas of Southern Methodist University. The principles of estimation were derived in part from a freshman engineering problem posed by Professor Thomas Kincaid of Boston University.

I would like to thank my family, Roxanne, Rachel, and Arielle, for giving me the time and space to think about and write this book. I also appreciate Roxanne's inspiration and help in identifying examples of human/machine interfaces.

Dedicated to Roxanne, Rachel, and Arielle

Charles B. Fleddermann is a professor in the Department of Electrical and Computer Engineering at the University of New Mexico in Albuquerque, New Mexico. He is a third generation engineer—his grandfather was a civil engineer and father an aeronautical engineer—so "engineering was in my genetic makeup." The genesis of a book on engineering ethics was in the ABET requirement to incorporate ethics topics into the undergraduate engineering curriculum. "Our department decided to have a one-hour seminar course on engineering ethics, but there was no book suitable for such a course." Other texts were tried the first few times the course was offered, but none of them presented ethical theory, analysis, and problem solving in a readily accessible way. "I wanted to have a text which would be concise, yet would give the student the tools required to solve the ethical problems that they might encounter in their professional lives."

Reviewers

ESource benefited from a wealth of reviewers who on the series from its initial idea stage to its completion. Reviewers read manuscripts and contributed insightful comments that helped the authors write great books. We would like to thank everyone who helped us with this project.

Concept Document

Naeem Abdurrahman- University of Texas, Austin
Grant Baker- University of Alaska, Anchorage
Betty Barr- University of Houston
William Beckwith- Clemson University
Ramzi Bualuan- University of Notre Dame
Dale Calkins- University of Washington
Arthur Clausing- University of Illinois at Urbana-Champaign
John Glover- University of Houston
A.S. Hodel- Auburn University
Denise Jackson- University of Tennessee, Knoxville
Kathleen Kitto- Western Washington University
Terry Kohutek- Texas A&M University
Larry Richards- University of Virginia
Avi Singhal- Arizona State University
Joseph Wujek- University of California, Berkeley
Mandochehr Zoghi- University of Dayton

Books

Stephen Allan- Utah State University
Naeem Abdurrahman - University of Texas Austin
Anil Bajaj- Purdue University
Grant Baker - University of Alaska - Anchorage
Betty Barr - University of Houston

William Beckwith - Clemson University
Haym Benaroya- Rutgers University
Tom Bledsaw- ITT Technical Institute
Tom Bryson- University of Missouri, Rolla
Ramzi Bualuan - University of Notre Dame
Dan Budny- Purdue University
Dale Calkins - University of Washington
Arthur Clausing - University of Illinois
James Devine- University of South Florida
Patrick Fitzhorn - Colorado State University
Dale Elifrits- University of Missouri, Rolla
Frank Gerlitz - Washtenaw College
John Glover - University of Houston
John Graham - University of North Carolina-Charlotte
Malcom Heimer - Florida International University
A.S. Hodel - Auburn University
Vern Johnson- University of Arizona
Kathleen Kitto - Western Washington University
Robert Montgomery- Purdue University
Mark Nagurka- Marquette University
Ramarathnam Narasimhan- University of Miami
Larry Richards - University of Virginia
Marc H. Richman - Brown University
Avi Singhal-Arizona State University
Tim Sykes- Houston Community College
Thomas Hill- SUNY at Buffalo
Michael S. Wells - Tennessee Tech University
Joseph Wujek - University of California - Berkeley
Edward Young- University of South Carolina
Mandochehr Zoghi - University of Dayton

Contents

7 DATABASE MANAGEMENT WITHIN EXCEL 95

8 COLLABORATING WITH OTHER ENGINEERS 105

9 EXCEL AND THE WORLD WIDE WEB 119

1

Engineering and Electronic Worksheets

1.1 INTRODUCTION TO WORKSHEETS

A *spreadsheet* is a rectangular grid composed of addressable units called *cells*. A cell may contain numerical data, textual data, macros, or formulas. Spreadsheet application programs were originally intended to be used for financial calculations. The original electronic spreadsheets resembled the paper spreadsheets of an accountant. One characteristic of electronic spreadsheets that gives them power over their paper counterparts is the ability to automatically recalculate all dependent values whenever a parameter is changed. Over time, more and more functions— including graphing functions, database functions, and the ability to access the World Wide Web—have been added to spreadsheet application programs. A large number of analytical tools—for example, scientific and engineering tools, statistical tools, data-mapping tools, and financial analysis tools—are now available within spreadsheet applications.

As an engineering student, you may find that an advanced spreadsheet program, such as Microsoft Excel, will suffice for many of your computational and presentation needs. For example, Excel may be used to manage small databases. If you wish to manage large or sophisticated databases, however, a specialized database application, such as Microsoft Access or Oracle, is preferable. You can use the Analysis Toolpack in Excel to perform mathematical analysis. If the analysis is large or sophisticated, however, you may want to use a specialized mathematical package, such as MathCAD or MATLAB. The same concept is true for graphing (Harvard Graphics) and statistical analysis (SPSS).

Microsoft Excel uses the term *worksheet* to denote a spreadsheet. A worksheet can contain more items than a

OBJECTIVES

After reading this chapter, you should be able to:

- Understand that the worksheet is an interface for engineering computation
- Understand the steps of the engineering method
- Understand engineering design and computers
- Understand how to use this book

traditional spreadsheet. Examples of these items are charts, links to Web pages, Visual Basic modules, and macros. We will treat the two terms synonymously in this text. A collection of worksheets that is stored in a single file is called a *workbook*.

1.2   THE ENGINEERING METHOD

Engineers are problem solvers. The ability to solve technical problems is both an art and a science. Solving of engineering problems requires a broad background in a variety of technical areas, such as mathematics, physics, and computational science. Solving of engineering problems also requires a set of nontechnical skills. Common sense and good judgement are examples of important nontechnical abilities. Engineering solutions often involve balancing several competing factors.

An example of such a trade-off is cost vs. reliability. For example, it costs money to remove impurities from an integrated circuit (IC) during the manufacturing process. Impurities are directly related to the reliability of the chip. As the manufacturing and testing process is improved to produce an increasingly reliable chip, the costs continue to rise (probably at a higher-than-linear rate). A design decision could be made that balances these two competing factors. We could express it in this way: A chip is reliable if there is a 90% probability that the chip will not fail in five years.

Another essential skill for engineers is the ability to collect and analyze data. We must be able to organize and communicate our results both verbally and in written form. The analysis and presentation of data using Microsoft Excel is the core topic of this book. In order to understand the role and importance of data analysis in the engineering profession, let us first look at a variation of a problem-solving process called the *engineering method*. The engineering method consists of a series of steps that help an engineer break the solution of a problem into smaller, logical phases. This method has been described many times in slightly different terms. One version of the method's steps is:

- Problem Definition
- Information Gathering
- Selection of the Appropriate Theory or Methodology
- Collection of Simplifying Assumptions
- Solution and Refinement
- Testing and Verification

Each of these steps will be described in the following sections. Within each of the six steps, an engineer typically analyses or presents some form of data.

1.2.1   Problem Definition

The precise definition of a problem can be one of the most difficult phases of solving a problem. Often a client (or instructor) will present a problem with ambiguous specifications. At times, the client does not completely understand the problem. It is worth the effort to obtain missing requirements and reduce ambiguities as early as possible in the design process. A clear and precise written statement of a problem at this stage will save much wasted expenditure of energy later.

1.2.2   Information Gathering

To solve a properly specified problem, the engineer must gather relevant information. At a minimum, this involves a thorough review of the relevant technical areas, this including a review of previous solutions to similar problems. The engineer may also perform experiments to test underlying assumptions that are needed before the solution to a problem can be devised. The experiments may involve laboratory testing or may use computer modeling.

For example, you might be asked to improve the drilling technology used at a particular offshore drilling site. The first thing to do would be to review reports of previous solutions used at this site and other sites. Perhaps you would obtain and analyze core samples from different depths at the drilling site. You might also take seismic readings and perform computer modeling using these data.

The engineer must be able to effectively communicate the results of testing and data collection. This often involves the use of tables and charts, as described in Chapter 5.

1.2.3 Selection of the Appropriate Theory or Methodology

A variety of scientific principles may be used to formulate the solution to a problem. The engineer's educational background and training contribute strongly to the ability to succeed in this step. However, after a theory or set of principles is chosen, it must be presented to others. Scientific principles are often expressed using mathematics. Chapter 4 describes how to create and use formulas within a spreadsheet.

1.2.4 Collection of Simplifying Assumptions

A theory is an abstraction of how the world works. To solve real-world problems, often make simplifying assumptions about some of the things the theory is about. As an example, suppose that you were given the job of designing a kiln. Part of the design process involves calculating the heat loss from the kiln. If you make the assumption that the convective loss is negligible, then the calculation only has to account for heat loss due to conduction and radiation. Excel provides a number of tools for viewing and analyzing data from different perspectives. These tools, discussed in Chapter 6, include trend analysis, goal seeking, pivot tables, and maps.

1.2.5 Solution and Refinement

Engineering problems are frequently solved iteratively. A common scenario for an engineer is to test and refine potential solutions to a problem using Excel. After the engineer is satisfied that the solution works for small data sets, the solution may be translated to a programming language, such as C or FORTRAN. The resulting program can then be executed on a powerful workstation or a supercomputer using large data sets. This use of a worksheet is called building a *prototype*. A spreadsheet package such as Excel is useful for building prototypes because solutions can be quickly developed and easily modified.

1.2.6 Testing and Verification

Final testing and verification of the solution to a problem is a critical step before the solution is implemented. Many times, misplaced decimal points or incorrect conversion of units produces unreasonable answers. Solutions can also be tested by choosing the endpoints of parameter ranges. In this way, the limits of the solution are tested. In many cases, it is impossible to test all of the possible input sets. For example, consider a control unit with eight input controls, each of which has 10 calibration settings. There are 100 million possible input combinations for such a control unit. It would be unreasonable to test every combination, so we choose a representative sample and present our results statistically. In Chapter 6, we discuss the statistical functions available in the Analysis Toolpack.

1.3 ISSUES IN ENGINEERING AND DATA ANALYSIS

1.3.1 Group Problem Solving

Engineers rarely work alone into today's environment. Teams of engineers work together in every stage of the design and problem-solving processes. The ability to work well in a group is an important skill to learn. One group task is the solution of engineering problems in a collaborative manner. When working in a group, it is necessary to incorporate each group member's input without inadvertently destroying the work of another member. It is also necessary to maintain historical versions of the group's work. In Chapter 8, we discuss some of the tools that Excel provides to assist with collaboration, including means for tracking changes, sharing a workbook, and protecting a workbook.

1.3.2 Document Transfer

During the group design process, data may be manipulated by a variety of application programs. Issues of data transfer and functional compatibility are more important than in the past. Prior to the 1990s, the design process generally occurred within a single organization that was located at a single geographical site. All engineers who worked on a project tended to use the same software and hardware. Today, members of design teams are likely to work from different geographical locations and may use different application programs and different computer architectures. The elements of a design project may pass hands many times during the design process. The ability to move data among software applications, operating systems, and hardware is important. The methods for importing and exporting data between Excel and other application programs are discussed in Chapter 8. The use of the Internet has become a major method of communicating ideas, transferring documents, and publishing information. The use of the Internet with Excel is discussed in Chapter 9.

1.3.3 Standardized Units

Since engineers from different sites and organizations tend to communicate and work more than ever as a team, it is important that the technical language they speak be the same. This technical language consists of mathematical notation and a system of units. The International System of Units (SI) is designed to be the basis for a worldwide standard for measurement. As an engineering student, you should become familiar with the SI system and with the methods for converting units among SI and other systems.

1.4 HOW TO USE THIS BOOK

This book is intended get the engineering student up and running with Excel 97 as quickly as possible. Examples are geared towards engineering and mathematical problems. Read the book while sitting in front of a computer. Learn to use Excel by re-creating each example in the text. Perform the instructions in the boxes labeled **Practice!** . Many of the worksheets used to develop the examples in the book are available for download via anonymous file transfer protocol (FTP) from

> **ftp.eng.fsu.edu/pub/kuncick/excel/**

Or point your Web browser to

> **ftp://ftp.eng.fsu.edu/pub/kuncick/excel/**

Chapter 9 discusses how to import worksheets from an FTP site directly into a local worksheet.

This book is not intended to be a complete reference manual for Excel. It is much too short for that purpose. Many books on the market are more appropriate for use as a complete reference manual. However, if you are sitting at the computer, one of the best reference manuals is at your fingertips. The on-line Excel help tools provide an excellent resource if properly used. Chapter 2 covers the use of on-line help.

1.4.1 Typographic Conventions Used in the Book

Throughout the text, the following conventions will be used:

Selection with the Mouse.   The book frequently asks you to move the mouse cursor over a particular item and then click and release the left mouse button. These actions are repeated so many times in the text that they will be abbreviated as

Choose **Item**.

If the mouse button is not to be released or if the right mouse button is to be used, then this will be stated explicitly.

Multiple Selections.   The book frequently refers to selections that require more than one step. For example, to see a print preview, perform the following steps:

1. Choose **File** from the Menu bar (at the top of the screen).
2. Choose **Print Preview** from the drop-down menu that appears after you perform step 1.

Multiple selections such as this will be abbreviated by separating choices with a comma. For example, the preceding two steps will be denoted as

Choose **File**, **Print Preview** from the Menu bar.

Multiple Keystrokes.   If you are asked to press multiple keys note that the book separates the keys with a plus sign. For example, to undo a typing change, you can simultaneously press the **Ctrl** key and the **Z** key. This will be described as

To undo typing, press **Ctrl** + **Z**. To redo typing, press **Ctrl** + **Y**.

Literal Expressions.   Underlines indicate a word or phrase that is a literal transcription. For example, the title bar at the top of the screen should contain the text <u>Microsoft Excel</u>.

Key Terms.   The first time a key term is used, it is italicized.

KEY TERMS

cells	selection of theory or methodology
collection of simplifying assumptions	solution and refinement
engineering method	spreadsheet
information gathering	testing and verification
problem definition	workbook
prototype	worksheet

Problems

1. Visit the U.S. National Institute of Science and Technology (NIST) Physics Laboratory's Web site about the International System of Units (SI) at

    http://physics.nist.gov/Divisions/Div840/SI.html

    Click on the menu item labeled <u>In-depth Information on the SI, the Modern Metric System</u>, and locate the table for SI Base Units. Use the table to fill in the missing entries in Table 1-1.

    TABLE 1-1   SI Base Units

QUANTITY	NAME	SYMBOL
length		m
	kilogram	kg
time	second	
electric current	ampere	
temperature		K
	mole	mol
luminous intensity		cd

2. The electronic spreadsheet has played an important role in the history of computing. The following links discuss the history of electronic spreadsheets. Access the following Web sites with your web browser, and then answer the following questions.

    Power, Daniel. *A Brief History of Spreadsheets,* at URL http://dss.cba.uni.edu/dss/sshistory.html, visited May 12, 1998.

    Browne, Christopher. *Historical Background on Spreadsheets,* at URL http://www.ntlug.org/~cbbrowne/spreadsheets02.html, visited May 12, 1998.

    Mattessich, Richard. *Early History of the Spreadsheet,* at URL http://www.j-walk.com/ss/history/spreadsh.htm, visited May 12, 1998.

    a. What was the name of the first commercially available electronic spreadsheet that was partly responsible for the early success of the Apple computer?

    b. What spreadsheet application remains the most widely sold software application in the world (as of 1998)?

2

Microsoft Excel Basics

2.1 UNDERSTANDING THE EXCEL SCREEN

This chapter introduces you to Microsoft Excel. The chapter is written for a student who is not familiar with the package, and it is designed to get you up and running as soon as possible. The preferred way to use this chapter is for you to sit at a computer and execute each of the examples as you read the chapter.

To start the Microsoft Excel program, place the cursor over the Excel icon, and click the left mouse button. The icon will resemble ⬚ or ⬚. The icon may be on your desktop or may be accessed from the Task bar if you are using Windows 95 or NT. A screen that resembles Figure 2.1 should appear. If the Tip of the Day box appears, close it for now by choosing **Close**. We will return to the Tip of the Day later in the chapter. Try to become familiar with each of the components on this screen, we will discuss since we will use these names throughout the book. Working generally from top to bottom, each of the components in turn.

2.1.1 Title Bar

The *Title bar* displays the Excel icon and the name of the worksheet currently being edited. Since we did not specifically open a worksheet, Excel supplies a default name (in our example, Book1). You should see several buttons on the right-hand side of the Title bar. These are used to manipulate the window and will be discussed later in chapter under the section Manipulating Windows. Figure 2.2 shows the Title bar from the sample screen in Figure 2.1.

2.1.2 Menu Bar

The *Menu bar* contains a list of menus. (See Figure 2.3.) If you place the cursor over an item on the Menu bar and click

OBJECTIVES

After reading this chapter, you should be able to:

- Understand the basic Excel screen layout
- Use the toolbars
- Access on-line help using various methods
- Create a new worksheet
- Open an existing worksheet
- Navigate through and edit a worksheet
- Print a worksheet

Materials science is an important field of study for engineers that covers the electronic, optical, mechanical, chemical, and magnetic properties of metals, polymers, composites, and ceramics.

Crystalline materials, such as metals and alloys, are subject to slip deformation when a shear strain is applied to the material. The deformation occurs when atomic planes slide along the directions of densest atomic packing. The following figure depicts slip deformation as being similar to the movement of a deck of cards under shear strain.

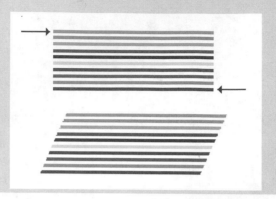

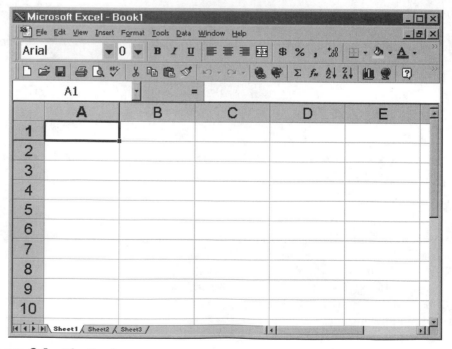

Figure 2.1. The Excel screen

Figure 2.2. The Title bar

the left mouse button, then a drop-down menu will appear. When you place the cursor over an item, the item should change color. While holding the cursor over an item, click the left mouse button to execute the item. Try the following steps:

1. Place the cursor over the **Window** menu item on the Menu bar.
2. Place the mouse over the item labeled **Arrange**. Click the left mouse button.

This book will use the following convention as a shortcut to denote the preceding steps:

From the Menu bar, choose **Window**, **Arrange**.

Figure 2.3.   The Menu bar

2.1.3 Toolbars

The next several rows of icons contain various *toolbars*. A toolbar is a group of buttons that are related to a particular topic (e.g., drawing). The use of buttons on toolbars is an alternative method of executing commands. Most (but not all) of the commands that can be executed from a toolbar can also be executed from the Menu bar. There are more than a dozen toolbars for various functions.

Figure 2.4 shows the toolbars from the example screen in Figure 2.1. The toolbars on your screen may not match these exactly. This is because the location and presence of toolbars may be customized. If all of the toolbars were displayed at once on the screen, there would likely not be much room left for anything else!

Figure 2.4.   Example toolbars

The toolbar in Figure 2.5 is called the Standard toolbar. It contains some of the most frequently used commands. If this toolbar does not appear on your screen, then choose **View**, **Toolbars** from the Menu bar, and check the box labeled <u>Standard</u>. This will place the Standard toolbar on your screen.

Figure 2.5.   The Standard toolbar

Move the mouse over the toolbar button that looks like 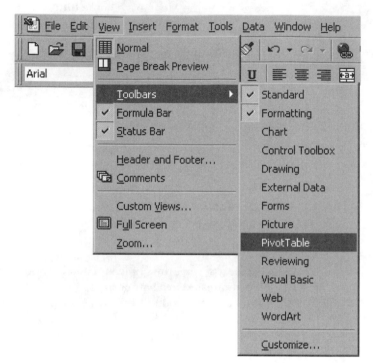, but don't click the mouse button. The word <u>Print Preview</u> should appear. Clicking the left mouse button on the icon has the same effect as choosing **File**, **Print Preview** from the Menu bar.

The choice and location of toolbars may be customized. To add or remove toolbars from the screen, choose **View**, **Toolbars** from the Menu bar. A list of toolbars that resembles Figure 2.6 should appear on your screen. Check any toolbars that you would like to have displayed on the screen.

Figure 2.6. Adding or deleting toolbars

PRACTICE!

Move the mouse slowly over each button on the Standard toolbar without pressing the mouse button. For example, move the mouse cursor over the icon. The word Paste should appear.

Try to locate the equivalent command on the Menu bar. Do this for each of the buttons on the Standard toolbar. Are there any buttons that don't have an equivalent Menu bar item? What about the icon?

Once a toolbar appears on the screen, it may be moved by dragging the toolbar with the mouse. This is accomplished by moving the cursor to the left side of the toolbar (there are two vertical lines on the left side of each toolbar), clicking the left mouse button, and using the mouse to drag the toolbar to a new location. Toolbars may be placed along the left-hand side of the screen or in a separate box anywhere on the screen. Figure 2.7 shows the Standard toolbar placed along the left side of the screen, the Formatting toolbar placed at the top of the screen, and the PivotTable toolbar placed near the center of the screen.

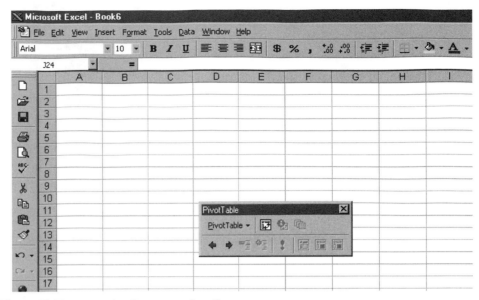

Figure 2.7. Example of customized toolbars

2.1.4 Formula Bar

The *Formula bar* displays the formula or constant value for the selected cell. Figure 2.8 shows an example of a Formula bar, which appears immediately below the Formatting toolbar. The selected cell is B5, and the formula for cell B5 is =SUM(B3,B4). If the Formula bar does not appear on your screen, then choose **View** from the Menu bar and check the box labeled Formula Bar. The maximum length for a formula is 1,024 characters.

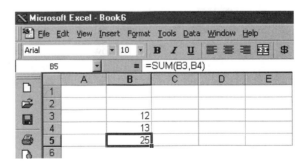

Figure 2.8. The Formula bar

2.1.5 Workbook Window

The workbook window is the area on the screen where data are entered. The maximum size for a worksheet is 65,536 rows by 256 columns. The columns are labeled A, B, C, … AA, AB, … IV, and the rows are labeled 1, 2, 3, … 65,536. An entire column can be selected by clicking the left mouse button on the column label. An entire row can be selected by clicking on the row label. The entire worksheet can be selected by clicking on the blank box in the top left corner of the workbook window. Figure 2.9 depicts a workbook window with column C selected.

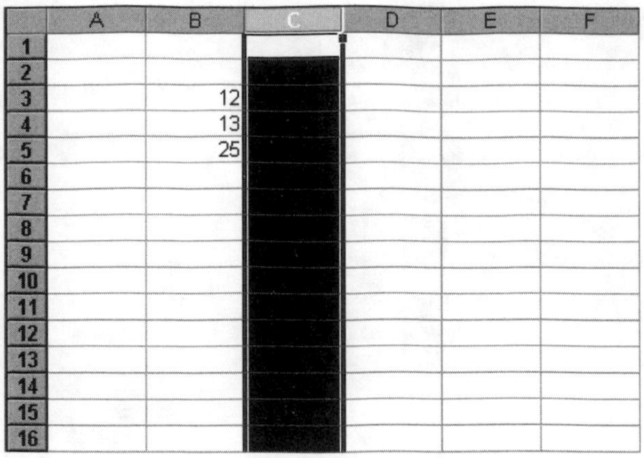

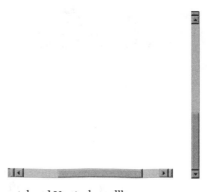

Figure 2.9. The Workbook window

2.1.6 Sheet Tabs

The Sheet Tab bar is positioned at the bottom of the screen. The Sheet Tab bar lists the worksheets in the workbook. (See Figure 2.10.) You can move quickly from sheet to sheet by selecting a sheet tab. If there are more sheets that are visible on the Sheet Tab bar, then you can use the arrows to the left of the sheet tabs to move from sheet to sheet. By default, Excel creates three worksheets when you create a new workbook. There is a maximum of 255 sheets in a default workbook.

Figure 2.10. Sheet tabs

2.1.7 Scrollbars

The Vertical and Horizontal *scrollbars* are located along the right-hand side and bottom of the Workbook window. Many worksheets are much larger than the visible window. A scrollbar makes it possible to move quickly to any position on a worksheet. You can move around the worksheet by dragging a scrollbar, or you can click on the arrows at either end of the scrollbar. Figure 2.11 shows the Horizontal and Vertical scrollbars.

Figure 2.11. The Horizontal and Vertical scrollbars

2.1.8 Status Bar

The Status bar is normally positioned at the very bottom of the Excel screen. The Status bar displays information about a command in progress and displays the status of certain keys, such as **Num Lock, Caps Lock,** and **Scroll Lock.** The Status bar is depicted in Figure 2.12 as showing the **Num Lock** key turned on. If the status bar is not visible on your screen, choose **View, Status Bar** from the Menu bar.

Figure 2.12. The Status bar

2.1.9 Shortcut Keys

Another method of executing Excel commands uses combinations of keys on the keyboard and bypasses the mouse altogether. For example, to execute the Copy and Paste functions using these combinations, or shortcut keys, first select a cell by placing the mouse cursor over the cell. Simultaneously hold down the **Ctrl** key and the key for the letter **C**. Throughout the book, the plus sign will denote keys that are to be pressed simultaneously; for example, the previous sentence will be worded, <u>Press the **Ctrl** + **C** keys</u>.

Using the shortcut keys to copy has the same effect as choosing the button from the Standard toolbar (or choosing **Edit, Copy** from the Menu bar). To paste the contents of the cell to a new location, first select a new cell in which to copy by moving the mouse cursor. Next, press **Ctrl** + **V**. Using this shortcut to copy has the same effect as choosing the button from the Standard toolbar (or choosing **Edit, Paste** from the Menu bar).

Now you know three ways of executing the Cut and Paste functions. Why does Excel have so many ways of performing the same function? The reason is that users with different levels of experience have different needs. The novice user may have trouble finding commands. Using the Menu bar is a good method for the novice, because the command names are listed and the commands are usually in the same place. As the novice gains experience, the toolbars become more useful, since a toolbar button is faster to execute than a Menu bar selection. As a user becomes very proficient with Excel and learns to type at a rapid rate, the shortcut keys become the quickest way to execute a command. Movement of the fingers from keyboard to the mouse is avoided. The downside of using keyboard shortcuts is that they have to be memorized.

This book will teach you only a few keyboard shortcuts. As you become more proficient at Excel, you might consider learning and memorizing shortcuts for several of your most frequently used commands. One method of learning them is to look at the right-hand side of the Menu bar items. For example, choose **Edit** from the Menu bar, and note that the shortcut for **Cut** listed on the menu is **Ctrl** + **X**.

2.2 GETTING HELP

Excel contains a large on-line help system. To access the help menu, choose **Help** from the Menu bar. There is a variety of ways to obtain help, including the following:

- Choosing a section from the table of contents
- Searching an index
- Searching for keywords in the help text
- Using the Office Assistant

- Using the What's This feature
- Learning from the Tip of the Day
- Accessing help from the World Wide Web
- Accessing special help for Lotus 1-2-3 users

Each of these methods will be discussed in the sections that follow.

2.2.1 Choosing from the Table of Contents

Choosing from the Table of Contents is useful if you have time to read about a general topic. Reading through the topics could serve as a tutorial. This is not the method to use if you have a specific question and you want an immediate answer. To access the Table of Contents, choose **Help**, **Contents and Index** from the Menu bar, and select the **Contents** tab. A good first selection for you to read is titled **Getting Help**.

2.2.2 Searching the Help Index

To access the Help Index, choose **Help**, **Contents and Index** from the Menu bar, and select the **Index** tab. A dialog box that resembles Figure 2.13 should appear.

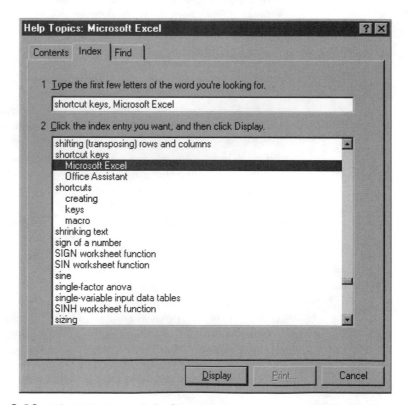

Figure 2.13. The Help Topics dialog box

Type in a key word or words, and choose a topic from the list that is displayed. For example, type shortcut keys and select **Microsoft Excel** from the list. Now choose the **Display** button. A help dialog box that resembles Figure 2.14 will appear. Choose any 🔲 button to see information about a selected topic. If you would like to print a help topic, then choose **Options**, **Print Topic** from the Help dialog box.

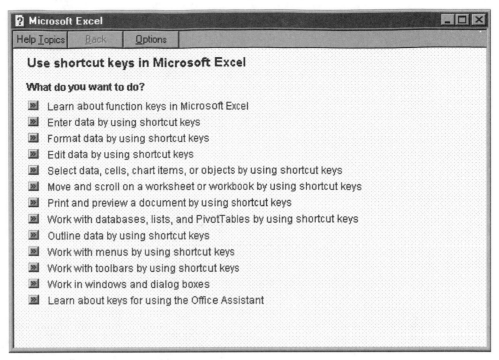

Figure 2.14. A Help dialog box

2.2.3 Searching for Keywords in the Help Text

This option provides a more exhaustive search capability by building a database of all words in the help text. Before the search can proceed, the database must be built. Fortunately, you have to build the list only once. To access the keyword search feature, choose **Help**, **Contents and Index** from the Menu bar, and then select the **Find** tab. A dialog box will appear. Perform the steps listed in the box:

1. Type in a keyword.
2. Select a matching word to narrow the search.
3. Select a topic from the list.

Figure 2.15 demonstrates the use of the keyword search feature to locate help about hyperlinks.

2.2.4 Using the Office Assistant

The Office Assistant is a new feature of Office 97 products. This feature is intended to interactively guide you through many tasks. The Office Assistant has several personalities, including an animated paper clip and an animated Albert Einstein. To change the personality of the Office Assistant, choose the ⑦ button on the Standard toolbar. When the Office Assistant dialog box appears, choose **Options**, **Gallery**. The `Next >` and `< Back` buttons can be used to view the various Office Assistant personalities. The Office Assistant has a search feature similar to the keyword and index searches previously mentioned. As you gain proficiency with Excel, you may reach a point where you no longer want to use the Office Assistant. To remove the Office Assistant, choose the `● Close` button on the Office Assistant picture.

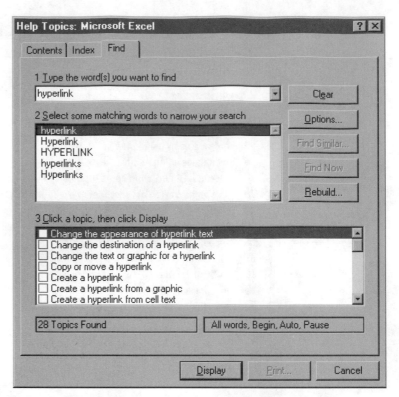

Figure 2.15.   Example of keyword search

2.2.5 Using the <u>What's This</u> Feature

If you would like to learn about a button or other graphic item on the screen, then the What's This feature may be helpful. Choose **Help**, **What's This** from the Menu bar. Your cursor should change to the shape ▶?. Now move the mouse to the icon or graphic item that you want to learn about and click the left mouse button. A help box will appear, and your cursor will return to its original shape.

2.2.6 Learning from the Tip of the Day

If you would like Excel to provide you with a helpful tip each time you start up the Word program, then open the Office Assistant by choosing the ? icon from the Standard toolbar. Next, choose **Options**, and select the **Options** tab. Click the item labeled **Show the Tip of the Day at Startup**

2.2.7 Accessing Help from the World Wide Web

A wealth of information about Microsoft Excel is available on the World Wide Web. To access an Excel-related Web site, choose **Help**, **Microsoft on the Web** and then select one of the items on the drop-down menu. For example, select **Free Stuff**, and you will be connected to a Microsoft site where you can download a new Office Assistant personality.

If these links do not work, then you may not be connected to the Internet. To access the World Wide Web, you must have access to the Internet through your school, work, or a private *Internet service provider* (ISP).

2.2.8 Accessing Special Help for Lotus 1-2-3 Users

If you are a Lotus 1-2-3 user, then you can access a help section that explains the differences between Lotus 1-2-3 commands and Excel commands. Choose **Help**, **Lotus 1-2-3 Help** from the Menu bar.

2.3 MANIPULATING WINDOWS

It is a good idea to spend some time getting used to manipulating windows before we actually begin to create a workbook. Excel allows you to keep more than one workbook open at a time. This is helpful when you are copying text or objects from one worksheet to another. There are three important buttons on the title bar of every workbook, and the same three buttons appear on the main Excel title bar. These buttons are used to control the window. The buttons and their functions are explained in Figure 2-1.

TABLE 2-1   The Window control buttons

BUTTON	NAME
▬	Minimize Control button
⊡	Full Screen Control button
✕	Close Control button

2.3.1 Minimize Control Button

The *Minimize Control button* reduces the window to an icon in the main Excel window. The name of the workbook is displayed on the icon. Figure 2.16 depicts three workbooks. Book 6 is open and Books 7 and 8 are minimized. To restore a minimized document, double click the worksheet's icon. When minimized, the document is not closed, and it remains in computer memory. Minimization is useful if you plan to use the document again shortly, since it will not have to be retrieved from disk.

2.3.2 Full Screen Control Button

The *Full Screen Control button* allows you to maximize the amount of "real estate" that your document or application occupies on the screen. Choose the ⊡ icon on the title bar to expand the document or application to full screen size. Choose the ⊡ icon to reduce a full-screen document or application to a smaller window size.

2.3.3 Close Control Button

The *Close Control Button* will close your workbook or application. If there are unsaved changes, then an alert box will appear that reminds you to save your work. The alert box is depicted in Figure 2.17. If you choose **Yes**, then the workbook or application will be saved and closed. If you choose **No**, then the workbook or application will be closed without saving. If you choose **Cancel**, then the workbook or application will not be closed, and you will be returned to it.

2.4 CREATING AND SAVING WORKSHEETS AND WORKBOOKS

2.4.1 Introduction to Templates

A *template* is a workbook that has some of its cells filled in. If you use similar formatting for many documents, then you will benefit from creating and using a template. You may

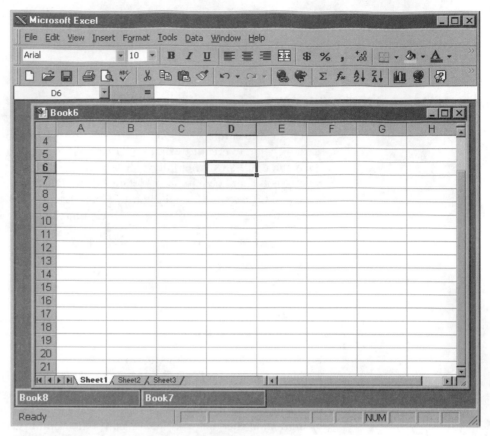

Figure 2.16. Example of minimized workbooks

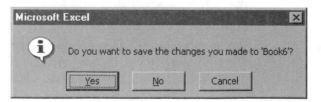

Figure 2.17. Alert box for saving changes

build your own template or customize preformatted templates and, in time, create a library of your own template styles.

2.4.2 Creating and Opening Documents

To create a new document, choose **File**, **New** from the Menu bar. The New dialog box depicted in Figure 2.18 should appear. The tabs on this dialog box refer to several groups of templates. Choose one of the tabs and click on a template. For example, choose the **Spreadsheet Solutions** tab and then select **Invoice**. Once you have selected a template, click **OK** to create the new document. As you can see, much of the tedious formatting used to create an invoice has already been done for you.

A quick method of creating a new blank workbook to click on the ⧠ button on the Standard toolbar. You will not be asked to choose from the template list.

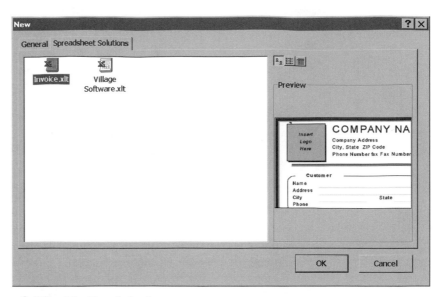

Figure 2.18.   The New dialog box

To open an existing document, choose **File**, **Open** from the Menu bar or choose the 📂 button on the Standard toolbar. The Open dialog box depicted in Figure 2.19 should appear. From this dialog box, you can type in a path and file name, or you can browse the file system to locate a file. If the file system is very large, then you may want to use the search function to locate files of a given name or files that contain a certain string of text.

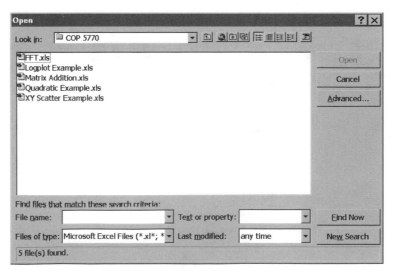

Figure 2.19.   The Open dialog box

2.4.3 Naming Documents
It is important to develop a methodical and consistent method for naming documents. Over time, the number of documents that you maintain will grow larger, and it will

become harder to locate or keep track them. Documents that are related should be grouped together in a separate folder. Do not use the default workbook names, e.g., Book1, Book2, Book3, etc., or chaos will soon ensue. If documents are not given meaningful names, then the documents may be inadvertently overwritten. Documents that have very general names, e.g., <u>Spreadsheet</u>, will be difficult to locate later.

By default, Excel documents are given the extension xls. Unless you are specifically creating a template (.xlt), ASCII text document (.txt), or other special type of document, you should use the default extension.

2.4.4 Opening Workbooks with Macros

A *macro* is a recording of a group of tasks that are stored in a Visual Basic module. A set of frequently repeated commands can be stored in a macro and then executed with a single mouse click whenever they are needed. Macros are very powerful tools. However, they can contain viruses that infect files on your computer. For this reason, you should enable macros only if you are certain of the origin of the document. For example, if you followed the previous example and opened the invoice template, a warning box should have appeared that resembles Figure 2.20. Since this template came with Excel, you may assume that it is safe and click Enable Macros.

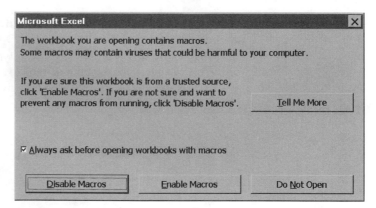

Figure 2.20.   The Macro Warning dialog box

2.4.5 Saving Documents

To save a document for the first time, choose **File**, **Save As** from the Menu bar. The Save As dialog box as depicted in Figure 2.21 should appear. Choose a folder in which to save the document by selecting the ▼ button on the right side of the box labeled **Save In**. Then type in (or select) a name for your document.

As mentioned in the section on naming documents, it is wise to carefully choose a meaningful name for your document. Once a document has been given a name, it may be reopened and edited. To save an open document that already has a name, choose **File**, **Save** from the Menu bar, or choose the 🖫 button from the Standard toolbar. If you are unsure of the name of the current working document, you can view it in the title bar.

Saving documents on a frequent basis an important task. It is also important to make backup copies of your vital documents on floppy disks or some other physical device. There are many tales of woe from students (and professors) who have lost hours of work after a power failure.

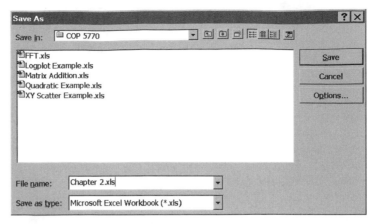

Figure 2.21. The Save As dialog box

Fortunately, Excel has several *AutoSave* features that make the frequent saving of documents an easy task. The task of making frequent backup copies to a different medium, e.g., floppy disk or tape, is something that you must perform yourself, however.

To set the AutoSave features, choose **Tools**, **AutoSave**. (If the AutoSave option does not appear on the Tools menu, then you can add it to the menu by choosing **Tools**, **Add-Ins** and then checking the **AutoSave** box and clicking **OK**). The AutoSave dialog box depicted in Figure 2.22 should appear. From the AutoSave dialog box, you can choose any or all of the following features:

- How often to AutoSave (10 minutes is a good starting choice)
- Whether to save all open workbooks or only the active workbook (the latter is a good starting choice)
- Whether you should be prompted before saving (the prompt can quickly get annoying, so you may want to "uncheck" this item)

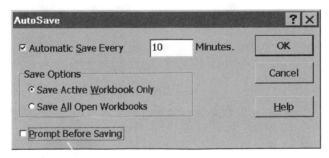

Figure 2.22. The AutoSave dialog box

2.5 EDITING A WORKSHEET

In addition to AutoSave, many other options may be set. To view the current settings, choose **Tools**, **Options** from the Menu bar and browse through the various tabs. We recommend that you leave the default settings for now if you are a new Excel user.

2.5.1 Moving Around a Worksheet

There are several methods of moving from place to place in an Excel worksheet. If the worksheet is relatively small, any of these methods will work equally well. As a worksheet grows in size, movement becomes more difficult, and you can save a lot of time by learning the various methods. The current cell number is displayed in the Name box on the left-hand side of the Formula bar. Figure 2.23 shows that H5 is currently the active cell.

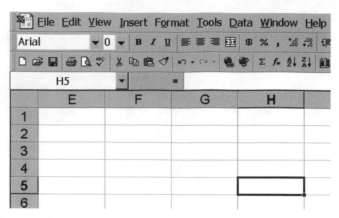

Figure 2.23.   The Name box

The general methods for moving around a document are as follows:

- Movement using the keyboard
- Movement using the mouse
- Movement using the Go To dialog box

Movement Using the Keyboard.   The keyboard may be used to select a worksheet from a workbook. The keyboard may also be used to navigate around a single worksheet quickly and effectively. There are many key combinations for moving through a worksheet and among a number of worksheets. Table 2-2 lists the most frequently used combinations for moving within a worksheet.

TABLE 2-2   Movement within a Worksheet Using the Keyboard

KEY COMBINATION	ACTION
Ctrl + Page Down	Select next worksheet
Ctrl + Page Up	Select previous worksheet
⇦	Move one cell (column) to the left
⇨	Move one cell (column) to the right
⇧	Move up one cell (row)
⇩	Move down one cell (row)
Page Down	Move down one window
Page Up	Move up one window
Ctrl + ⇨	Move to right column of worksheet
Ctrl + ⇩	Move to bottom row of worksheet

Movement Using the Mouse.   The mouse may be used to select a worksheet and to move within a worksheet. To select a worksheet, choose a tab from the Sheet Tab bar depicted in Figure 2.10.

One method of moving around a worksheet with the mouse is to click on a cell. This is most useful if the new insertion point is located on the same screen. If the desired location is on a different page, then the scrollbars and scrolling arrows may be used to move quickly to a distant location. Figure 2.19 shows the vertical scrollbar and arrows.

Figure 2.24.   The Vertical scroll bar and arrows

Movement using the Go To dialog box.   Open the Go To dialog box by choosing **Edit**, **Go To** from the Menu bar, or press the **F5** key. The <u>Go To</u> dialog box depicted in Figure 2.25 will appear. A history of previous references is kept in the <u>Go To</u> window so that recently visited cells can be quickly located.

Click the **Special** button on the <u>Go To</u> dialog box, and the <u>Go To</u> Special dialog box depicted in Figure 2.26 should appear. You can select items from this dialog box to locate and browse a particular type of item. For example, Figure 2.26 shows that only cells containing formulas are to be located.

2.5.2 Selecting a Region

Much of the time spent preparing a spreadsheet preparation involves moving, copying, and deleting regions of cells, or other objects. In this section, we will be selecting regions of cells but the same principles apply to regions that contain charts, formulas, and other objects. Before an action can be applied to a region, the region must be selected. The selection process can be performed by using either the mouse or the keyboard.

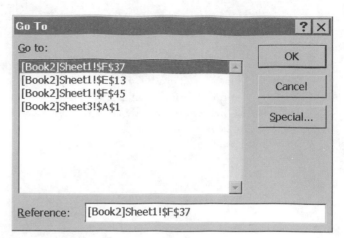

Figure 2.25. The Go To dialog box

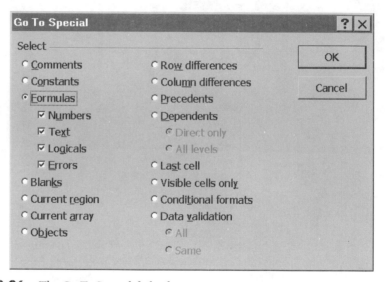

Figure 2.26. The Go To Special dialog box

To select a region of text with the mouse, move the mouse cursor to the beginning of the region, click, and drag to the end of the region. As you drag the mouse, the selected region will be highlighted.

To select a region of text that is larger than one screen, move the mouse cursor to the bottom of the screen. If you hold the mouse and leave the cursor at the bottom of the screen without releasing the mouse button, then the selected region will continue to grow, and the screen will scroll downward. This may take a little practice.

An alternative method for selecting large regions of a document is the following:

1. Move the mouse cursor onto one corner of the region that you wish to select.
2. Hold down the **Shift** key, and scroll to the ending location using one of the arrow keys, the **Page Up** key, or the **Page Down** key.

To select the entire worksheet, choose the **Select All** button at the top left corner of the worksheet. The Select All button is depicted in Figure 2.27. This is useful if you are applying a change to every cell in a worksheet.

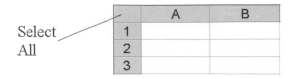

Figure 2.27. The Select All button

If you make a mistake and incorrectly select a region, then click the mouse cursor anywhere on the document window before you apply an action (such as delete). If the highlighting disappears, then you have deselected the region.

PRACTICE!

Try the following exercise to practice selecting regions:

1. Place the cursor over cell B3 and press **5**.
2. Press the down arrow key.
3. Press **6**.
4. Press the down arrow key.
5. Press **7**.
6. With the mouse, place the cursor over cell B5, hold down the left mouse button, and drag the mouse until cells B5, B4, and B3 are all highlighted.
7. Now choose the AutoSum button Σ on the Standard toolbar.

A new cell will be added that contains the sum of cells B3, B4, and B5. The results should resemble Figure 2.28.

	A	B
1		
2		.
3		5
4		6
5		7
6		18

Figure 2.28. The AutoSum feature

2.5.3 Cutting, Moving, Copying, and Pasting

Once a region has been selected, several actions may be taken, such as delete, move, copy, and paste. As usual, there is a variety of ways to accomplish these actions. One method uses the right mouse button, which we have not yet used.

Cutting a Region.   Cutting a region places it on the clipboard. To cut a region by using the mouse, first select the region by one of the methods described earlier. Then choose **Edit**, **Cut** from the Menu bar. The cut region will be highlighted by a rotating dashed line. An alternative method is to click the right mouse button and choose **Cut** from the Quick Edit menu that appears. The Quick Edit menu is depicted in Figure 2.29.

Figure 2.29.   The Quick Edit menu

Moving a Region.   A region may be moved by first cutting it and then pasting it to a new location. Try this by selecting a region, clicking the right mouse button, and choosing **Cut** from the Quick Edit Menu. Now place the insertion point in a new location, and either

- create a region of the same size and shape as the cut region, or
- select a single cell.

Click the right mouse button, and choose **Paste** from the Quick Edit Menu. The original region of cells should now appear in the new location. If you do not create the new region with the same size and shape, then an error box will appear to prompt you to do so.

An advantage of the cut-and-paste method of moving a region is that the region may be moved across documents and even across applications. You can cut a region from an Excel worksheet and paste that region into a Word document.

An alternative method for moving a region is to select the region and click and hold down the left mouse button anywhere on the edge of the region. Now drag the region to the new location. When the selected region is in the correct location, release the mouse button.

Copying a Region.   Copying a region is similar to moving it, except that the original copy of the region remains intact. To copy a region, first select the region to be copied. Then choose **Edit**, **Copy** from the Menu bar. Place the insertion point in the new location, and choose **Edit**, **Paste** from the Menu bar to make a copy. The last action can be performed as many times as needed if multiple copies are to be made.

An alternative method of copying a region is to click the right mouse button and use the Quick Edit menu to perform the same actions. As mentioned in the section on moving a region, the copy-and-paste method can be used to copy regions to another document or another application.

2.5.4 Inserting and Deleting Cells

New cells may be added to a worksheet, and cells may be deleted. To delete a region of cells, first select the region. Then choose **Edit**, **Delete** from the Menu bar. An alternative method is to click the right mouse button and select **Delete** from the Quick Edit menu. In either case, the Delete dialog box depicted in Figure 2.30 should appear.

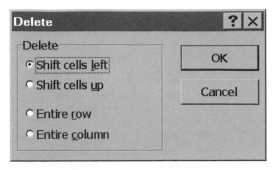

Figure 2.30. The Delete dialog box

Choose the direction to shift the remaining cells in the worksheet. If you want to clear the contents of a region of cells without shifting the other cells, then click the right mouse button and choose **Clear Contents** from the Quick Edit menu.

You can insert new cells, rows, columns, or an entire worksheet by selecting **Insert** from the Menu bar. The Insert drop-down menu is depicted in Figure 2.31.

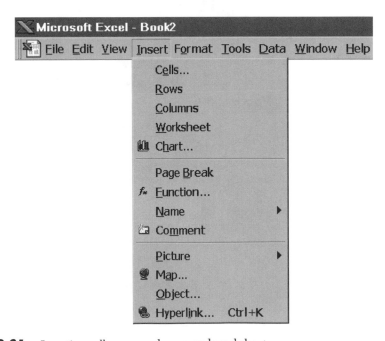

Figure 2.31. Inserting cells, rows, columns, and worksheets

2.5.5 Undoing Mistakes

Excel allows actions to be undone or reversed. To undo the last action, choose **Edit**, **Undo** from the Menu bar or type **Ctrl + Z**. To see the list of recent actions, choose the down arrow button ▾ next to the undo button ↶ on the Standard toolbar. From this list, you may select one or more actions to be undone. Note that if you select an action on the list, then all of the actions above it will also be undone! If you accidentally undo an action, then you may redo it by selecting the redo button ↷ on the Standard toolbar.

2.5.6 Checking Spelling

Excel can check the spelling of cells containing text. To check the spelling in a region, first select the region and then choose **Tools**, **Spelling**. Alternative methods are to press the **F7** key or choose ✓ from the Standard toolbar. If Excel finds a spelling mistake, then the Spelling dialog box depicted in Figure 2.32 will appear.

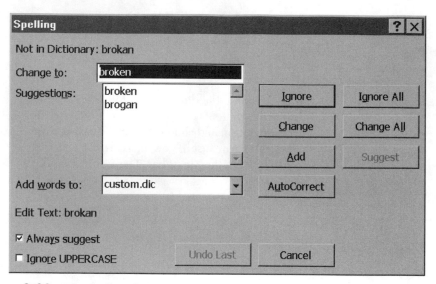

Figure 2.32. The Spelling dialog box

The region containing the mistake is displayed in the top text box. Suggestions for changes are presented in the bottom text box. At any point in the process, you can choose whether to accept or ignore the suggestions. If you choose a suggested correction, then you may click the **Change All** button to change all occurrences of the misspelled word in the selected region.

You may add new words to the main dictionary by choosing the **Add** button. This will probably be necessary as you proceed through your coursework, since many engineering terms are not in the custom dictionary.

2.5.7 The AutoCorrect Feature

The Excel *AutoCorrect* feature recognizes spelling errors and corrects them automatically. AutoCorrect performs actions such as automatically capitalizing the first letter of a sentence and correcting a word whose first two letters are capitalized. You can test to see if the AutoCorrect feature is turned on for your installation of Excel by typing the letters yuo and pressing the spacebar. Was the word automatically retyped as you? If so, then you have AutoCorrect turned on. To see the AutoCorrect options and dictionary, choose **Tools**, **AutoCorrect**. The AutoCorrect dialog box depicted in Figure 2.33 will appear.

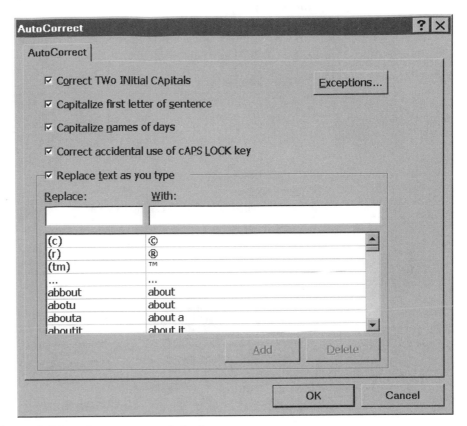

Figure 2.33. The AutoCorrect dialog box

From this dialog box, you can select (or deselect) various AutoCorrect options. You can also scroll through the AutoCorrect dictionary, add entries to the dictionary, and add exceptions to the dictionary. This last feature is necessary for exceptions to the selected options. For example, if you have selected the option that automatically converts the second capital letter to lowercase, you may have an occasional exception. One example is the abbreviation for the chemical compound hydrochloric acid, HCl. Another is the abbreviation of megahertz to MHz.

Be careful when adding new entries into the AutoCorrect dictionary. You may inadvertently add an entry for a misspelling that is a legitimate word.

2.6 PREVIEWING AND PRINTING A WORKSHEET

Before attempting to print a document, make sure that your printer is configured correctly. See your operating system documentation for assistance.

2.6.1 Previewing a Worksheet

It is advisable to preview a document before printing it. Many formatting problems can be resolved during the preview process, and trees will be saved. First, select a region to print. Then choose **File**, **Print Area**, and **Set Print Area** from the Menu bar. The selected region will now be surrounded by a dashed line.

To preview the document as it will be printed, select **File**, **Print Preview** from the Menu bar or choose the ◻ button from the Standard toolbar. The selected region will be displayed in the same format in which it will be printed. In addition, the Print Preview toolbar is placed on the screen. The Print Preview menu bar is displayed in Figure 2.34. The available options on the Print Preview menu bar are listed in Table 2-3.

TABLE 2-3 Print Preview toolbar options

BUTTON	ACTION
Next	Display next page of the worksheet
Previous	Display previous page of the worksheet
Zoom	Toggle between magnified and normal view
Print	Print the worksheet
Setup	Set the page orientation, margins, page order, etc.
Margins	Graphically set the margins and page stops
Page Break	PreviewGraphically set the page breaks
Close	Close current window and return to the worksheet
Help	Special help for the Print Preview menu bar

Figure 2.34. The Print Preview menu bar

2.6.2 Printing a Worksheet

To print a document, choose **File**, **Print** from the Menu bar. The Print dialog box depicted in Figure 2.35 will appear. To send a job directly to the printer without going through the Print dialog box, select the ▤ button on the Standard toolbar. The Print dialog box contains several choices, including commands for collating, selecting the number of copies to print, and selecting a range of tickets. A user may save a great deal of paper by using the Print Preview feature and selecting and printing only those pages that have been modified.

SUMMARY

This chapter introduced Microsoft Excel. The basic Excel components, including the Title bar, Menu bar, scroll-bars, and various toolbars were presented. Several methods for accessing on-line help were demonstrated. The chapter showed how to create a new worksheet and presented the basic commands for editing and printing a worksheet.

KEY TERMS

AutoSave
Close Control button
Formula bar
Full Screen Control button
Internet Service Provider
macro

Menu bar
Minimize Control button
Status bar
template
Title bar

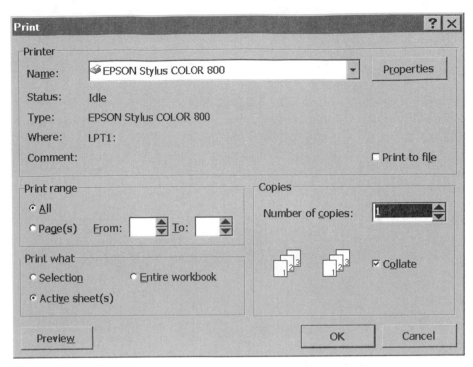

Figure 2.35. The Print dialog box

Problems

1. Practice cutting and pasting regions of cells. Can you cut and paste a region of Excel cells into a Word document? What about into an MS-DOS window?

2. As you type, make notes of your most common misspellings. After you have collected a list, add the misspelled words to the AutoCorrect dictionary.

3. Excel is a large and complex application. To see the variety of commands and tools that are available, browse through the various toolbars. Choose **View**, **Toolbars**, **Customize** from the Menu bar. The Customize Toolbar dialog box will appear. Check each of the 22 tool-boxes listed in the dialog box. As the toolbars appear on your screen, slowly drag your mouse cursor across the toolbar icons, and view the drop-down titles that appear. Figure 2.36 shows the Trace Dependents icon on the Auditing toolbar.

Figure 2.36. The Auditing toolbar

4. Familiarize yourself with the large volume of on-line help. Choose **Help**, **Contents and Index** from the Menu bar. The Help Topics dialog box will appear. Select, in turn, the **Contents**, **Index**, and **Find** tabs, and scroll through the list that appears in each case.

3

Entering and Formatting Data

3.1 ENTERING DATA

Cells can be filled with numerical values, text, times, dates, logical values, and formulas. In addition, a cell may contain an error value if Excel cannot evaluate its contents. The use of formulas is covered in Chapter 4. The other types of cell values are covered in this section.

3.1.1 Numerical Data

Numerical values containing any of the following symbols can be entered into a cell:

```
0 1 2 3 4 5 6 7 8 9
+ - ( ) , /
$ % .
E e
```

Although numerical values are stored internally with up to 15 digits of precision (including the decimal point), they can be displayed in a variety of formats. To see a list of cell formats, first type a number into a cell, and then click the right mouse button. The Quick Edit menu depicted in Figure 3-1 will appear.

Select **Format Cells** from the Quick Edit menu. The Format Cells dialog box depicted in Figure 3-2 will appear. Choose the **Number** tab. From this dialog box, the numerical value that you entered may be formatted as currency, a date, or a fraction, or you may even format it in scientific notation. When you change the format, the internal representation is not changed; only the method for displaying the value is modified. If the *General format* is selected, Excel will attempt to choose a format based on the contents of the cell. For example, if you type $3.4 into a cell, Excel will

SECTIONS

- 3.1 Entering Data
- 3.2 Formatting Worksheets
- Summary
- Key Terms

OBJECTIVES

After reading this chapter, you should be able to:

- Enter various types of data into a worksheet
- Quickly enter series of data
- Format rows, columns, and cells

Engineering economics involves the study of interest, cash flow patterns, techniques for maximizing net value, depreciation, and inflation. This is an important area of study for all engineers, since engineers frequently serve as managers or executive officers of corporations. The table that follows depicts the accumulated value of $10,000 that is invested by two students when they are 18 years old. The first student deposits the money in a savings account with 6% interest. The second student invests in the stock market, which averages an 11% annual gain. The second student retires comfortably at age 68.

By the time you have finished this chapter, you will be able to create and format tables such as this one. In Chapter 3, you will learn how to create formulas such as the one that computed the interest in the table.

Accumulated Capital		
Age	6% Growth	11% Growth
18	$10,000.00	$10,000.00
28	$17,908.48	$28,394.21
38	$32,071.35	$80,623.12
48	$57,434.91	$228,922.97
58	$102,857.18	$650,008.67
68	$184,201.54	$1,845,648.27

Figure 3.1.   The Quick Edit menu

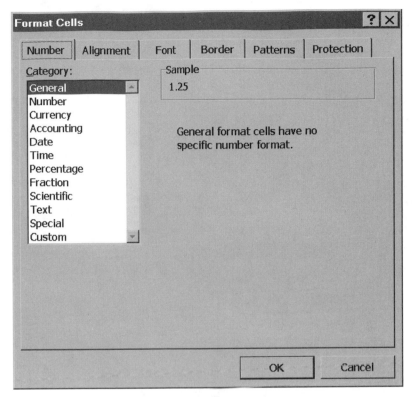

Figure 3.2. The Format Cells dialog box

automatically convert to *Currency format,* and the value will be right justified and displayed as $3.40.

3.1.2 Text Data

To treat the contents of a cell as text, first select the cell and then apply the text format. Numerical values that are entered into a text cell cannot be used for calculation. You can perform many of the same functions or the contents of a text cell that are used for a Word document. These include changing the font size, changing the font type, and checking spelling.

Once a cell has been stored internally as a number, it is slightly more difficult to convert it to an internal text representation. To do so, first select the cell, then change the format to text, and finally, press **F2** then **Enter**.

3.1.3 Date and Time Data

Excel stores dates and times internally as numbers allowing you to perform arithmetic on them. For example, you can subtract one date from another. If a dash (-) or slash (/) is inserted between two digits, then Excel assumes that the number is a date. If a colon (:) is used to separate two digits, then Excel assumes that the number represents time. Excel also recognizes the key characters AM and A to represent A.M. and P.M. and P to represent P.M.

PRACTICE!

Try entering data using several formats. First, select cells A1 and A2. Apply the number format to these cells with three decimal places of precision. Then type in the values 1.25 and 2.45, respectively. The results should be right aligned and represented as 1.250 and 2.450, respectively.

Now select cell A3, choose the AutoSum Σ button from the Standard toolbar, and press **Enter**. The value 3.7 should appear in cell A3. Note that only one decimal place is shown. Since you did not specify a format, Excel assumed a general format and displayed only the significant digits.

Let's see how text formatting differs from numerical formats. Select cells C1 and C2. Apply the text format to these cells and then type in the values 1.25 and 2.450, respectively. The results should be left aligned, and the decimal places should appear exactly as you typed them.

Now select cell C4, then choose the AutoSum button Σ from the Standard toolbar, and press **Enter**. The following formula should appear in cell C4:

    =SUM()

Type C1:C2 inside the parentheses so that the formula looks like

    =SUM(C1:C2)

and press **Enter**. Cell C4 should now contain a zero! This is because Excel cannot sum text cells, so it returns a zero result. Your worksheet should resemble Figure 3-3.

	A	B	C
1	1.250		1.25
2	2.450		2.450
3	3.7		
4			0

Figure 3.3. Text and numerical formats

3.1.4 Fill Handles

Data entry can be tedious. The use of fill handles allows one to quickly copy a cell into a row or column of cells. Fill handles can also be used to create a series of numbers in a row or column. The *fill handle* appears as a small black square in the bottom right corner of a selected region. A fill handle is depicted in Figure 3-4. When the mouse cursor is placed over the fill handle, its shape will change to a black cross. Click and hold the right mouse button while dragging the mouse to the right so that the cursor is over four or five cells. When the mouse button is released, the value in the original cell will be copied into the new cells.

The fill handle can also be used to create a series of numbers, dates, or times. The series can be linear or exponential. To see the fill series options, place the mouse over a fill handle, and click the right mouse button and drag the mouse. When the mouse is released, the Fill Series drop-down menu depicted in Figure 3-5 will appear. The use of fill series for data analysis is discussed in Chapter 5.

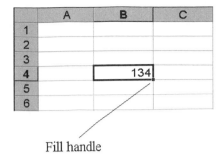

Figure 3.4. Example of a fill handle

Figure 3.5. The Fill Series drop-down menu

PRACTICE!

Practice using the fill handle by trying the following steps. First, try copying the contents of a cell. Type the value 1.5 into cell A1. Grab the fill handle with the right mouse button and drag it over cells B1:G1. When you release the mouse button, the value 1.5 should appear in cells A1:G1.

Now try creating a linear series. Type the values 1.5 and 2.5 into cells A3 and B3, respectively. Select the region A3:B3, and grab the fill handle with the right mouse button. Drag the fill handle over the cells C3:G3. Select **Linear Trend** from the drop-down menu.

Finally, create a growth series. Type the values 1.5 and 2.5 into cells A5 and B5, respectively. Select the region A5:B5, and grab the fill handle with the right mouse button. Drag the fill handle over the cells C5:G5. Select **Growth Trend** from the drop-down menu.

The results of your three exercises should resemble Figure 3-6.

	A	B	C	D	E	F	G
1	1.5	1.5	1.5	1.5	1.5	1.5	1.5
2							
3	1.5	2.5	3.5	4.5	5.5	6.5	7.5
4							
5	1.5	2.5	4.166667	6.944444	11.57407	19.29012	32.15021

Figure 3.6.   Examples using the fill handle

3.2 FORMATTING WORKSHEETS

3.2.1 Formatting Cells

In the previous section, you were shown how to format numerical data. Several other cosmetic formatting options are available for selected cells, including a choice of fonts, colors, borders, shading, and alignment.

To access the cell-formatting options, first select a region of cells and then choose **Format**, **Cells** from the Menu bar. An alternative method is to click the right mouse button and choose **Format Cells** from the Quick Edit menu. The format Cells dialog box depicted in Figure 3-7 will appear. From this dialog box, you may choose one of six tabs to format the font characteristics, alignment within the cell, borders, colors, fill patterns, and password protection of a worksheet.

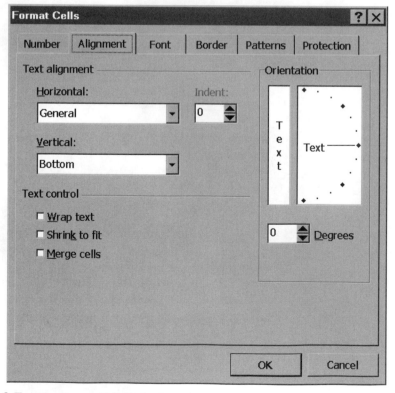

Figure 3.7.   The Format Cells dialog box

3.2.2 Formatting Columns and Rows

The primary use of the row-and-column formatting option is to determine the row height and the column width. You can specify the height and width exactly, or you can ask Excel to *AutoFit* the columns and rows for you. The AutoFit function adjusts the selected columns to the minimum width required to fit the data. If the data change then the AutoFit may have to be reapplied.

A second use of the row-and-column formatting option is to hide or unhide a row or column. This may be useful for simplifying the view of a complex worksheet. To view the column (or row) formatting options, first select a region, then choose **Format**, **Column** (or **Row**) from the Menu bar. The drop-down menu depicted in Figure 3-8 will appear.

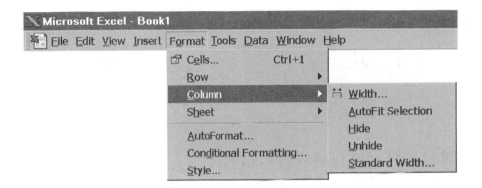

Figure 3.8. The Column Format drop-down menu

3.2.3 Table AutoFormats

Excel provides a few preformatted table types for convenience. To access the AutoFormat function, first select a region. Then choose **Format**, **AutoFormat** from the Menu bar. The AutoFormat dialog box depicted in Figure 3-9 will appear. Browse through the table formats in the list, and view the samples in the Sample box.

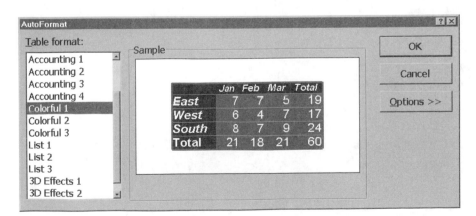

Figure 3.9. The AutoFormat dialog box

3.2.4 Formatting Entire Worksheets

Several formatting options apply to an entire worksheet. A worksheet can be hidden or renamed, or a different background can be selected. These functions can be accessed by choosing **Format**, **Sheet** from the Menu bar. A workbook can easily grow into a collection of dozens of worksheets. It is helpful to give worksheets meaningful names instead of using the default Sheet 1, Sheet 2, etc.

SUMMARY

This chapter introduced the data types used in Excel and the methods for entering data into a worksheet. The chapter also presented the commands for formatting cells, rows, columns, and worksheets.

KEY TERMS

AutoFit   fill handle
currency format   general format
engineering economics

Problems

1. A General format cell is interpreted by Excel. The application attempts to determine the numerical type from clues in the number. For example, if you type $3.56 into a cell, Excel guesses that you are representing money and changes the format type to Currency. What format type does Excel use if you enter a fraction such as 1/2 into a General format cell? What type does Excel use if you type 0 1/2?

2. Have some fun with the cell-formatting options. Play with the options under the **Font**, **Border**, **Alignment**, and **Patterns** tabs. If you get in trouble, learn to use the Undo feature by choosing Edit, Undo from the Menu bar.

3. Excel stores numbers with 15 digits of precision. Prove to yourself these limits of numerical precision. Select an empty cell and format the cell to the Number format with 20 decimal places. Select the equal sign on the Formula bar and type the formula

   ```
   =SQRT(2)
   ```

   Since the square root of two is an irrational number, the fractional part of its decimal representation has an unending number of nonrepeating digits. At what number of digits does Excel's accuracy stop?

4

Engineering Computation

4.1 INTRODUCTION

In this chapter, you are introduced to Excel's basic computational tools that are used to solve engineering problems.

The ability to manipulate formulas, arrays, and mathematical functions is the most important feature of Excel for engineers. A common scenario for an engineer is to test and refine potential solutions to a problem using Excel. After the engineer is satisfied that the solution works for small data sets, the solution may be translated to a programming language, such as C or FORTRAN. The resulting program can be then executed on a powerful workstation or supercomputer using large data sets. This use of a worksheet is called building a *prototype*. A spreadsheet package such as Excel is useful for building prototypes because solutions can be quickly developed and easily modified.

In the following sections, formulas and functions will be used to solve two problems that should be familiar to engineering students: (1) finding the solutions to a quadratic equation and (2) matrix multiplication. These examples will be used to demonstrate many of the features of Excel that are related to engineering computation.

4.2 CREATING AND USING FORMULAS

A formula in Excel consists of a mathematical expression. The cell containing the formula can display either the formula definition or the results of applying the formula.

The default is to display the results in the cell. This is usually preferable, since the formula definition for the currently active cell is displayed in the Formula bar.

Figure 4-1 shows cell D1 to be the currently active cell. The formula bar shows the formula definition of D1 to

OBJECTIVES

After reading this chapter, you should be able to:

- Create mathematical formulas in a worksheet
- Use Excel's predefined functions
- Debug worksheet formulas with errors
- Perform simple matrix operations with Excel

Electrical engineering is an important and diverse field that includes the study of circuit theory, electronics, electromagnetics, and the principles of power distribution. One small area is the study of direct-current (DC) circuits. At the moment a switch in a DC circuit is turned on or off, a *transient circuit* condition occurs.

A general expression for the current I in a DC transient circuit is

$$I(t) = I_\infty + \left(I_0 - I_\infty\right)e^{-t/T}.$$

At the end of this chapter, you will learn how to compute the current in a DC transient circuit using Excel.

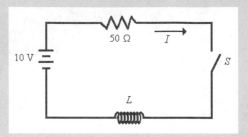

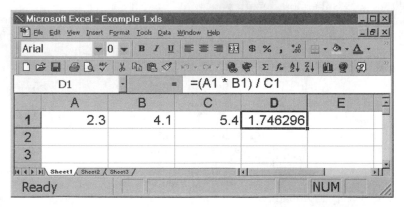

Figure 4.1. An example of a formula

be = **(A1 * B1) / C1.** The result of applying this expression is displayed in cell D1 as 1.746296. In the following paragraphs, we will show you how to build and debug formula definitions.

4.2.1 Formula Syntax

An Excel formula has a strict syntax. A formula can consist of operators, cell references, cell names, and function names. The use of each of these syntactic groups will be covered in this chapter.

A formula always begins with an equals sign (=). This symbol is an indicator for Excel to evaluate the expression that follows instead of simply placing the contents of the expression contents in the cell. (Try removing the equals sign and see what happens.) Since this is the most common error that users make when creating a formula, Microsoft has automated the placement of the equals sign when one uses the Formula Editor. The

Formula Editor can be accessed by clicking on the = button at the left end of the Formula bar. We recommend the use of the Formula Editor because the user is given immediate feedback about syntax and execution errors. A variety of predefined functions may be selected directly from the Formula Editor. (See the section on **Mathematical Functions** later in this chapter.) Table 4-1 displays the available arithmetic *operators*.

TABLE 4-1 Arithmetic operators

OPERATOR	OPERATION
%	Percentage
^	Exponentiation
°, /	Multiplication, Division
+/ -	Addition, Subtraction

The operators are listed in order of precedence. For example, exponentiation will be calculated before addition. If a different precedence is desired, then parentheses must be used. There are other operators for the manipulation of text and for Boolean comparison that will not be covered here. (See the on-line help section for further information.)

Cell references can be entered in two ways. A cell location can be typed into the formula editor, or the cell can be selected using the mouse. Let's walk through an example that uses both methods of referencing cells:

1. Enter the values 7.5 and 6.2 into cells A3 and B3, respectively.
2. Select cell C3.
3. Choose the = button. The Formula Editor will appear, and an equal sign will automatically be entered as the first character in the formula.
4. Select cell A3 by clicking on it with the left mouse button. A3 will be added to the formula in the Formula Editor.
5. Move the cursor to the formula in the Formula Editor and type

    A3 * B3 + A3^2

    The screen should resemble Figure 4-2. Note that the results of evaluating the formula are immediately displayed in the Formula Editor as you build the formula.
6. Click on **OK** and the Formula Editor will disappear. The formula result of **102.75** will now be displayed in cell C3.

4.2.2 Selecting Ranges

In the previous example, single cells were selected (or typed) into the formula definition. A range of cells can also be used in a formula. A range of cells can be entered by dragging the mouse cursor across the cells. Be careful not to insert a range into an expression that doesn't make sense mathematically. For example, the formula

> = SUM (D4:D8)

will sum cells D4 to D8. But the formula

> = SQRT (D4:D8)

makes no sense, since you can't take the square root of a range of numbers. When an invalid range is entered, Excel responds by placing the following error message in the target cell:

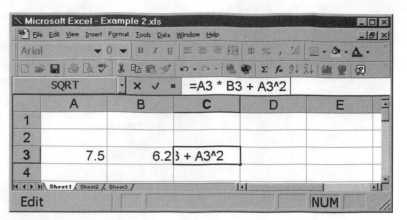

Figure 4.2.   Creating a formula

#VALUE.

4.2.3 Cell and Range Names

A group of cells can be named and the name added to a stored list. The name can then be used to reference the group of cells in a formula. Consider a range of cells that represents the following matrix:

	A	B	C
1	1	-1	2
2	4	0	-1
3	-8	2	-2

Figure 4.3.   A 3×3 matrix

This matrix can be represented in a worksheet, as depicted in Figure 4-3. Assume that in a formula you want to perform operations on the diagonal of the matrix. For example, the following formula will sum the diagonal elements:

= SUM (A1, B2,C3)

The following steps demonstrate how to name the diagonal elements:

1. Select the diagonal cells. Click on A1. Hold down the **Ctrl** key and click on B2. Hold down the **Ctrl** key and click on C3.
2. Choose Insert, Name, Define. A drop down-menu titled Define Name should appear.
3. Type in the name of the group of cells (e.g., **Diagonal**). Your screen should resemble Figure 4-4.
4. Click **OK** to finish the operation.

The name Diagonal is now associated with the cell range (A1, B2, C3) and can be used anywhere the range is referenced. To use a name in a formula, first place the cursor at the insertion point and then choose **Insert**, **Name**, **Paste**. A drop-down box with the list of names will be displayed. Names can also be typed directly into formulas.

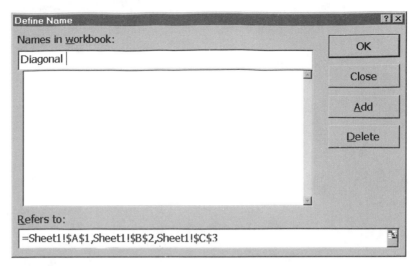

Figure 4.4. Naming a region of cells

For example, a formula that sums the diagonal elements of the matrix can now be stated as

= SUM (Diagonal)

This is more readable and less prone to typographical errors than the original version of the formula. The importance of using names will become clear as you begin to build more complex formulas (and attempt to debug errors in them).

4.2.4 Absolute and Relative References

Formulas may be copied from one location to another in a worksheet. Usually, you will want the referenced cells to follow the formula. For example, you may want to copy a formula that sums a row of numbers. Figure 4-5, notice that cell E4 contains a formula that sums the range A4:C4. If cell E4 is copied to cell E5 by using the **Edit**, **Copy**, and **Paste** selections, then cell E5 will contain a copy of the formula. Note that the range of cells that are summed has *followed* the formula. The formula in E5 sums A5:C5, not A4:C4!

E5		=	=SUM(A5:C5)	
	A	B	C	D	E
1					
2					
3					
4	3	4	7	Sum=	14
5	5	12	2	Sum=	19

Figure 4.5. A relative reference

This operation is called *relative referencing*. The range in the formula **= SUM (A4:C4)** is, by default, a relative reference. There are times when you may wish to copy a formula and not have the references follow the formula. This is called *absolute refer-*

encing. A cell or range is denoted as an absolute reference by placing a dollar sign ($) in front of the row or column to be locked.

In Figure 4-5, change the formula in E4 to read = **SUM ($A44:$C$4)**. Now copy the formula to cell E5. Notice that the cell references have not followed the formula and the resulting sum in cell E5 is still 14.

PRACTICE!

> There are times when you may want to make a relative reference to some cells and keep other references fixed. For example, a formula may use a constant and several variables.
>
> Place the following formula in cell C7 to compute the circumference of a circle
>
> = **$C$3 * B7^2**
>
> Create a worksheet that resembles Figure 4-6. Copy the formula to cells C8, C9, and C10. The reference to C3 (pi) should remain fixed, since it is an absolute reference, and the reference to B7 (radius) should follow the formula, since it is a relative reference.
>
> Create a formula that computes the area of a circle, and add it to your worksheet. Use an absolute reference for the constant *pi*. Use a relative reference for the variable *radius*.

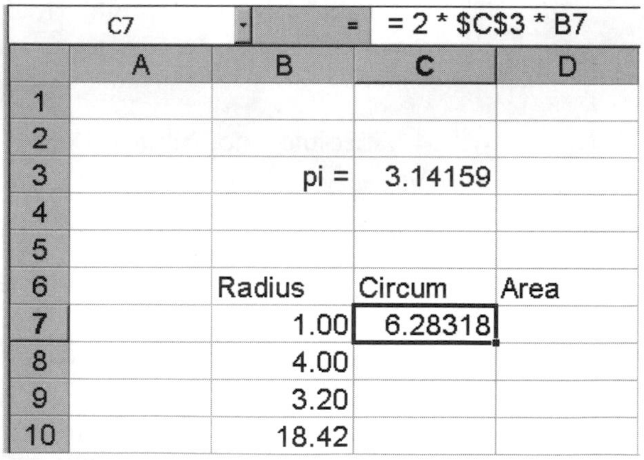

C7		=	= 2 * $C$3 * B7
	A	B	C	D
1				
2				
3		pi =	3.14159	
4				
5				
6		Radius	Circum	Area
7		1.00	6.28318	
8		4.00		
9		3.20		
10		18.42		

Figure 4.6. Mixed relative and absolute references

Example: Solution of Quadratic Equations   An example of a slightly more complex use of formulas is the solution of quadratic equations. You may recall from high school algebra that if a quadratic equation is expressed in the form

$$ax^2 + bx + c = 0,$$

then the solutions for x are

$$x = -b \pm \frac{\sqrt{b^2 - 4ac}}{2a}, \quad (a \neq 0).$$

Since Excel does not directly recognize imaginary numbers, we must make the further restriction that

$$b^2 - 4ac \geq 0.$$

Assume that a, b, and c are stored in cells A2, B2, and C2, respectively. We will place the formulas for the two solutions in D2 and E2, respectively. The Excel formula for the first solution is

> = (-B2 + SQRT (B2^2 - 4 * A2 * C2)) / (2 *, A2)

and the formula for the second solution is

> = (-B2 - SQRT (B2^2 - 4 * A2 * C2)) / (2 *. A2)

PRACTICE!

Use the quadratic solutions just given to practice entering and using formulas in Excel. Label the columns so that your worksheet resembles Figure 4-7. Copy the formulas so that you can enter up to five sets of coefficients. (Don't type them five times). Notice how the row numbers in the formulas change, since you are using relative references. Also, notice that an error is displayed for the solutions on row 5, since one of our assumptions ($a \neq 0$) is violated.

D6	·	=	= (-B6 + SQRT(B6^2 - 4 * A6 * C6)) / (2 * A6)

	A	B	C	D	E
1	Coefficient A	Coefficient B	Coefficient C	Root 1	Root 2
2	1	2	1	-1.0000	-1.0000
3	1	16	1	-0.0627	-15.9373
4	-4	-8	24	-3.6458	1.6458
5	0	18	24	#DIV/0!	#DIV/0!
6	-4	-8	4	-2.4142	0.4142

Figure 4.7. Computing solutions of a quadratic equation

4.2.5 Error Messages

The formulas that we have presented so far are relatively simple. If you make an error in typing one of the sample formulas, the location of the error is relatively easy to spot. As you begin to develop more complex formulas, locating and debugging errors becomes a more difficult problem. When a syntactic error occurs in a formula, Excel will attempt to immediately catch the error and then display an error box that explains the error. However, formulas can be syntactically correct, yet still produce errors upon execution. If an expression cannot be evaluated, Excel will denote the error by placing one of eight error messages in the target cell. These error messages are listed in Table 4-2.

4.2.6 Debugging Errors Using Cell Selection

Debugging errors in a worksheet is made much easier by the use of the special cell selection option. To view the special cell selection menu, first select a range of cells to view. Then click **Edit**, **Go To**, and **Special**. The Go To Special dialog box depicted in Figure 4-9 should appear.

From the cell section menu, a variety of options may be used to assist in the debugging process. We will discuss several of the options that are most relevant to debugging mathematical formulas.

TABLE 4-2   Excel error messages

######	The value is too wide to fit in the cell, or an attempt was made to display a negative date or time.
#VALUE	The wrong type of argument was used in a formula. (For example, text was entered when an array argument was expected.)
#DIV/0	An attempt was made to divide by zero. (See the quadratic equation example in Figure 4-7.)
#NAME	A name is not recognized. Usually, the function or defined name was misspelled. Note that named ranges or functions may not contain spaces.
#N/A	This error means "not available." It occurs most often when a function has been given an incorrect number of arguments or, in the case of matrix functions, when the argument matrix size does not correspond to the actual matrix size. Figure 4-8 demonstrates an attempt to place the inverse of a (2×2) matrix into a (3×3) matrix.
#REFA	A referenced cell is not valid. This usually occurs when a cell is referenced in a formula and then that cell is deleted. It also occurs if an attempt is made to paste a cell over a referenced cell.
#NUM	The expression produces a numeric value that is out of range or invalid. Examples are extremely small or large numbers, or imaginary numbers. Try the formula **= SQRT (-1)**
#NULL	An attempt was made to reference the intersection of two areas that do not intersect.

D2	▾	= {=MINVERSE(A2:B3)}		
	A	**B**	**C**	**D**	**E**
1					
2	3	4		-3	2
3	5	6		2.5	-1.5
4				#N/A	#N/A

Figure 4.8.   Example of a #N/A (not available) error

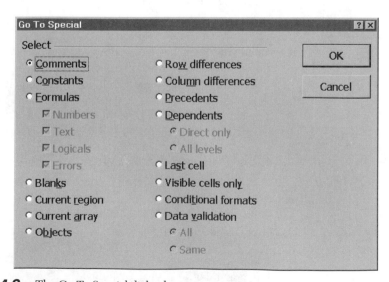

Figure 4.9.   The Go To Special dialog box

Formulas. A common worksheet error is the accidental replacement of a formula with a constant. The selection of the **Formula** option will highlight all cells that contain formulas. You can further refine this option by selecting formulas that result in numerical, text, logical, or error values.

Precedents. This option displays cells that precede the selected cell(s). The displayed cells are all necessary for the computation of the selected cell(s). For example, if cell D4 contains the formula = **A3 * B3** then cells A3 and B3 precede D4. The option may be refined by choosing either **Direct Only** or **All Levels**. The Direct Only option will display the immediate precedents. The All Levels option will recursively display precedents (precedents of precedents). If cell A3 contained the formula **=SUM(A1:A2)**, then A1 and A2 (as well as A3 and B3) would show up as precedents of D4 if the All Levels option is selected.

Dependents. This option displays cells that depend on the selected cell(s). For example, if you select cell B3 and choose the **Dependents** option, then cell D4 will be displayed. This is because the formula in cell D4 depends on B3 for its computation.

Column (or Row) Differences. This option highlights cells in a column or row that have a different *reference pattern* from the other cells in the column or row. For example, if all but one of the selected cells in a column contain formulas, and the exception contains a constant, then the exception will be displayed.

The following example demonstrates the use of the cell selection technique. Figure 4-10 illustrates a worksheet that computes the standard deviation of the numbers in column A. The difference between a data value and the mean value is computed in column B using the formula **=A1-$E$2** . This difference is squared in column C using the formula = **B1^2** . The problem is that the answer is incorrect! The standard deviation should be 13.67, not 13.02. The error was found by using the following steps:

1. Select column B by clicking on the column label.
2. Choose **Edit**, **Go To**, and **Special**.
3. Choose **Column Differences** and click **OK**.

The result is that cell B5 is highlighted. By looking in the formula bar, you can see that B5 incorrectly contains a constant value instead of a formula. The debugging tools prove their worth as the worksheet gets larger and more complex. (Consider trying to find the same error without the debugging tool if column A contained 3,000 rows.)

B5		=	-8.54				
	A	B	C	D	E	F	G
						Sum of Differences Squared	Standard Deviation
1	32	0.7143	0.5102		Mean		
2	14	-17.2857	298.7959		31.2857	1017.8500	13.02466
3	52	20.7143	429.0816				
4	26	-5.2857	27.9388				
5	18	-8.5400	72.9316				
6	45	13.7143	188.0816				
7	32	0.7143	0.5102				

Figure 4.10. Using cell selection to debug a formula

4.2.7 Debugging Errors Using Tracing

Excel also provides a visual method for tracing precedents, dependents, and cells with errors. The auditing tool may be accessed by choosing **Tools**, **Auditing**. The easiest way to manipulate the visual tool is to select **Show Auditing Toolbar** The Auditing toolbar contains buttons for tracing precedents, dependents, and cells with errors. The tool draws blue arrows showing the direction of precedence. Cells with arrows are boxed in red. Figure 4-11 demonstrates the effect of choosing **Trace Precedents** from the Auditing toolbar when cell F2 is selected. Every cell in column C is a precedent of the formula in cell F2 = **SUM (C1:C7)**.

	A	B	C	D	E	F	G
						Sum of Differences Squared	Standard Deviation
1	32	0.7143	0.5102		Mean		
2	14	-17.2857	298.7959		31.2857	1121.4286	13.67131
3	52	20.7143	429.0816				
4	26	-5.2857	27.9388				
5	18	-13.2857	176.5102				
6	45	13.7143	188.0816				
7	32	0.7143	0.5102				

Figure 4.11. The Trace Precedents feature

4.3 MATHEMATICAL FUNCTIONS

Excel has a large number of built-in functions that are similar to functions in a programming language. A function takes a specified number of arguments as input and returns a value. Excel 97 functions are organized into groups, including database functions, financial functions, text functions, and date/time functions. Three groups that are of the most interest to engineers are the mathematical functions, the logical functions, and the statistical functions. We will focus on the use of these three function groups.

To browse the available functions, click on the f_{x} Paste Function button on the Standard toolbar. The Paste Function dialog box depicted in Figure 4-12 will appear. You can use this dialog box to select and paste functions into a formula or merely to browse and learn about the available functions and their syntax.

The following steps will walk you through the use of two simple statistical functions that compute the mean and median of a list of numbers:

1. Enter seven midterm grades into the range A1:A7. (You can use the numbers from the previous example under **Debugging Errors**.) Format the cells to be numbers with one digit to the right of the decimal point.

2. Name the cell range (e.g., **MidTerm**) by choosing **Insert**, **Name**, **Define** from the Menu bar.

3. Select a cell to hold the mean (average) of the midterm grades, select the Paste Function key **f***.

4. The Paste Function dialog box will appear. Select **Statistical** from the Function Category list, and select **Average** from the Function Name list. Click **OK**.

5. The Formula Editor should now appear. Erase whatever is on the line labeled Number 1, and type **MidTerm**. The Formula Editor should resemble Figure 4-13.

6. Click OK to finish the operation.

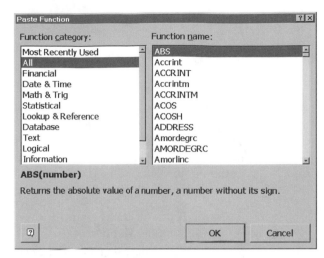

Figure 4.12. The Paste Function dialog box

7. Perform similar steps for the median. (The only difference is that you choose **Median** from the Function Name list.)

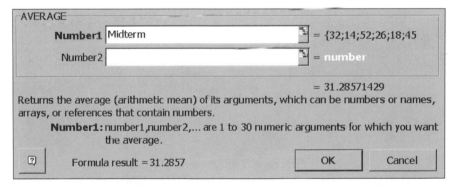

Figure 4.13. The Formula Editor

4.4 MATRIX OPERATIONS

Matrices, or arrays, are frequently used in formulating and solving of engineering problems. A *matrix* is defined to be a rectangular array of elements each of which is referenced by a row and a column number. A spreadsheet is a natural application to represent and manipulate matrices. Excel has a number of built-in matrix operations that are called *array functions*. These include:

MDETERM(array)– returns the matrix determinant for the named array.

MINVERSE(array)– returns the inverse of the named array.

MMULT(array1, array2)–performs matrix multiplication on the two named arrays.

There are also several predefined functions that compute sums or differences of products on matrices. In addition, many other functions take ranges as arguments and can be used to evaluate a matrix. For demonstration purposes, we'll use the two follow-ing matrices,which are entered in ranges A3:B4 and D3:E4, respectively:

$$A = \begin{bmatrix} 3 & 1 \\ 4 & 3 \end{bmatrix}. \quad B = \begin{bmatrix} 3 & -5 \\ 1 & 0 \end{bmatrix}.$$

The following steps will guide you through performing matrix addition. Recall that matrix addition is performed by adding each of the corresponding cells of two matrices. The two matrices must be of the same *order*, which means that they both have the same number of rows and same number of columns. The order of a matrix is often described as the number of rows by the number of columns (rows × columns). The given sample matrices are of order (2 × 2).

1. Name matrices *A* and *B* by choosing by choosing **Insert**, **Name**, **Define** from the Menu bar.
2. Select the range G3:H4 by clicking the mouse and using it to drag the cursor over these cells.
3. Type the following formula into the formula bar:

    = A + B

4. Simultaneously press the keys **Ctrl**, **Shift**, and **Enter**. Curly braces will appear, enclosing the formula. (*Note*: You cannot type the curly braces; the **Ctrl**, **Shift**, and **Enter** key sequence must be used.) The range G3:H4 will now display the matrix sum of *A* and *B*. Your screen should resemble Figure 4-14.

G3		= {=A + B}					
	A	B	C	D	E	F	G	H
1								
2								
3	3	1		3	-5		6	-4
4	4	3		1	0		5	3

Figure 4.14.   Example of matrix addition

As a second example, we will walk you through the execution of the *transpose* operation. The transpose of a matrix is the matrix that is formed by interchanging the rows and columns of the original matrix. To find the transpose of *B*:

1. Select a range of empty cells in which to store the results. The selected range must be of the same order as *B* (2 × 2).
2. Type the following into the formula bar:

    = TRANSPOSE(B)

3. Simultaneously press the keys **Ctrl**, **Shift**, and **Enter**. Curly braces will appear, enclosing the formula. The result should be

$$B^T = \begin{bmatrix} 3 & 1 \\ -5 & 0 \end{bmatrix}.$$

PRACTICE!

Matrix multiplication is defined as follows: If $A = \begin{bmatrix} a_{ij} \end{bmatrix}$ is an $m \times n$ matrix and $B = \begin{bmatrix} b_{ij} \end{bmatrix}$ is an $n \times p$ matrix, then the *product* $AB = C = \begin{bmatrix} c_{ij} \end{bmatrix}$ is an $m \times p$ matrix defined by

$$c_{ij} = \sum_{k=1}^{n} a_{ik} b_{kj}, \quad i = 1, 2, \ldots, m, \quad j = 1, 2, \ldots, p.$$

From the preceding example, the product of A and B is calculated to be

$$AB = \begin{bmatrix} 3 & 1 \\ 4 & 3 \end{bmatrix} \begin{bmatrix} 3 & -5 \\ 1 & 0 \end{bmatrix} = \begin{bmatrix} (3 \cdot 3) + (1 \cdot 1) & (3 \cdot -5) + (1 \cdot 0) \\ (4 \cdot 3) + (3 \cdot 1) & (4 \cdot -5) + (3 \cdot 0) \end{bmatrix} = \begin{bmatrix} 10 & -15 \\ 15 & 20 \end{bmatrix}.$$

Excel has a built-in matrix multiplication function named **MMULT()**. Use this function to verify the foregoing results. You can practice using this function even if you have not yet studied matrix multiplication.

APPLICATION: DC CIRCUITS

A general expression for the current I in a DC transient circuit is

$$I(t) = I_\infty + \left(I_0 - I_\infty \right) e^{-t/T},$$

where

I_0 is an initial value at the instant of sudden change,

I_∞ is the current at time $t = \infty$,

$T = RC$ is the time constant for a series R-C circuit, and

$T = L/R$ is the time constant for a series R-L circuit.

If the switch is closed at $t = 0$, then we can calculate the current in the circuit after three time constants ($t = 3T$). Since $I_0 = 0$ and $I_\infty = V/R$, a worksheet can be set up that calculates $I(3T)$ for various values of V and R. If I_0, V, and R are placed into cells B5, C5, and D5, respectively, then the Excel formula for is

= C5 / D5

and the Excel formula for is

= E5 + (B5 – E5) * EXP(F5)

If you enter the voltage (10 volts) and resistance (50 ohms) from the preceding diagram of a circuit, then your results should resemble Figure 4-15. The current at three time constants after the switch is closed equals 0.190043 A. Try entering various values for V and R.

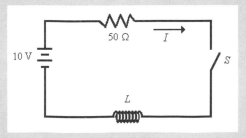

	A	B	C	D	E	F	G
1							
2							
3							
4		I(0)	V	R	I(infinity)	-t/T	I(3T)
5		0	10	50	0.2	-3	0.190043

Figure 4.15. Computing the current in a DC transient circuit

SUMMARY

The use of mathematical formulas and functions in worksheets is important for engineers. This chapter introduced and demonstrated the use of formulas and functions in Excel. Mathematical and engineering functions were emphasized and methods for debugging and auditing worksheets were presented. Matrix representation and manipulation are natural tasks for a spreadsheet application. Several matrix operations were demonstrated.

KEY TERMS

absolute reference	precedents
cell references	prototype
dependents	reference pattern
matrix	relative reference
operators	transient circuit
order	transpose

Problems

1. Create an Excel formula that will compute the following equation for $x = 1, 2, ..., 10$:

$$y = \ln x + e^x \sin x .$$

Write the formula once, and use the fill handle to drag the formula over the other nine cells.

2. The formula that calculates the number of combinations of r objects taken from a collection of n objects is

$$C(n, r) = \frac{n!}{(n - r)!\, r!} .$$

Thus, the number of collections of eight people that can fit into a six passenger vehicle is calculated as

$$C(8, 6) = \frac{8!}{(8 - 6)!\, 6!} = 28.$$

Write an Excel equation to calculate combinations.

3. Excel has a number of predefined logical functions. One of these, the IF() function, has the following syntax:

```
IF (EXP, T, F)
```

The effect of the function is to evaluate EXP, which must be a logical expression. If the expression is true, then T is returned. If the expression is false, then F is returned.
For example, the function

```
IF (X < 200, X, "Cholesterol is too high")
```

returns the value of X if X is less than 200. But, if X is greater than or equal to 200, then the text statement "Cholesterol is too high" is placed in the selected cell.

Expand the quadratic equation presented in this chapter to test for division by zero. If the expression $2a = 0$ is true, then display "Divide by Zero"; otherwise return the value of $2a$.

Perform a similar test for $b^2 - 4ac \geq 0$; display "Requires Complex Number" if the test is false.

4. The horizontal range of a projectile fired into the air at angle θ degrees is given by

$$R = \frac{2V^2 \sin \theta \cos \theta}{g} ,$$

assuming no air resistance. Create a worksheet that computes R for a selected initial velocity V and firing angle θ. Use g = 9.81 meters/sec². Convert degrees to radians using the RADI-

ANS() function. An initial velocity of 150 meters/sec and firing angle of 25° should result in $R = 1756$ meters.

5. Two frequently performed matrix operations are the calculation of the *determinant* of a matrix and the *inverse* of a matrix. Expand on the examples used in the chapter to find the determinant of matrix *A* and the inverse of matrix *B*.

5

Working with Charts

5.1 CREATING CHARTS

5.1.1 Using the Chart Wizard to Create an XY Scatter Chart

The Chart Wizard guides you through the construction of a chart. Once the chart is built, its components may be modified. Before proceeding, create the worksheet depicted in Figure 5-1. The data were collected by measuring the current (I) across a resistor for eight measured voltages (V). *Ohm's law*, $V = IR$, states that the relationship between V and I is linear if the temperature is kept relatively constant. An XY scatter plot of the data in Figure 5-1 can be used to visualize this relationship.

Before starting the Chart Wizard, select the region containing the data (A2:B9). Start the Chart Wizard by choosing the Chart Wizard button ▦ from the Standard toolbar, or alternatively, choose **Insert**, **Chart** from the Menu bar. The first Chart Wizard dialog box, depicted in Figure 5-2 should appear. This dialog box prompts you to choose a chart type. Select **XY (Scatter)** from the list labeled Chart type. Select the top left box from the area labeled Chart sub-type, and choose the   button to proceed.

The second Chart Wizard dialog box should now appear. Choose the **Series** tab, as depicted in Figure 5-3. Excel has chosen, by default, to plot cells (A2:A9) on the X axis, and cells (B2:B9) on the Y axis.

You can modify the selected regions for the X values or the Y values by choosing the ▦ button on the right end of the boxes labeled X Values or Y Values. The Source Data dialog box depicted in Figure 5-4 will appear. The currently selected region will be surrounded by a dashed line. If you

OBJECTIVES

After reading this chapter, you should be able to:

- Understand the principles used to build all of the different types Excel charts
- Know the specific methods used to build line charts and XY scatter plots
- Understand the available options for editing and formatting chart legends, axes, and titles
- Scale axes and create error bars

Fluid mechanics deals with the study of fluids in both the gaseous and liquid state. The study of fluid mechanics includes statics, kinematics, and the dynamics of fluids. Some of the common properties of fluids are temperature, density, specific weight, viscosity, specific gravity, and the speed of sound within the fluid.

The temperature of the atmosphere varies with altitude above the earth. The following chart displays the atmospheric temperature for selected altitudes. By the time you have finished studying this chapter, you will be able to create and format charts using Excel.

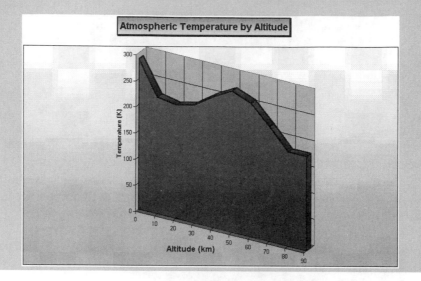

	A	B
1	Potential (V)	Current (A)
2	6.97	0.051
3	5.96	0.044
4	4.95	0.038
5	3.98	0.032
6	3.03	0.025
7	1.91	0.018
8	1.02	0.012
9	0.5	0.008

Figure 5.1. Data collected by measuring current across a 150-Ω resistor

wish to change the X values, use the mouse to select a region, and then choose the 🔲 button from the Source Data dialog box to return to the Chart Wizard. For our current example, the X values do not need to be modified. Choose Next > to proceed.

The third Chart Wizard dialog box, depicted in Figure 5-5, should appear. This dialog box guides you through the chart formatting options. From the third Chart Wizard dialog box, you can create and modify the chart titles, axes, grid lines, data labels, and the legend. Choose the **Titles** tab, and type the chart title, X axis, and Y axis titles as depicted in Figure 5-5. Remove the legend by choosing the **Legend** tab. Make sure that the box titled Show Legend is not checked. Choose Next > to proceed to the fourth and final step.

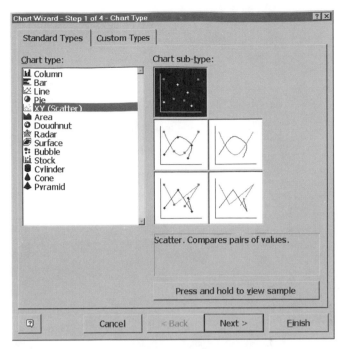

Figure 5.2. The Chart Wizard dialog box: Step 1

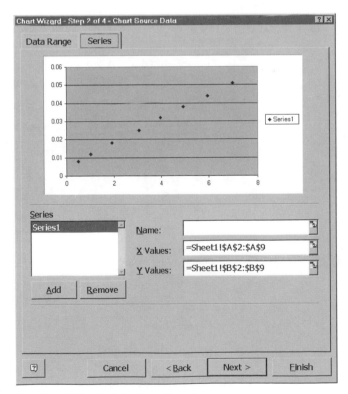

Figure 5.3. The Chart Wizard dialog box: Step 2

	A	B
1	Potential (V)	Current (A)
2	6.97	0.051
3	5.96	0.044
4	4.95	0.038
5	3.98	0.032
6	3.03	0.025
7	1.91	0.018
8	1.02	0.012
9	0.5	0.008

Chart Wizard - Step 2 of 4 - Chart Source Data - X Values:
=Sheet1!$A$2:$A$9

Figure 5.4. Choosing a region for X values in a chart

Figure 5.5. The Chart Wizard dialog box: Step 3

The fourth Chart Wizard dialog box, depicted in Figure 5-6 should now appear. This dialog box prompts you to choose a location for the chart. There are two choices.

Figure 5.6. The Chart Wizard dialog box: Step 4

The first choice, titled <u>As new sheet</u>, will place the chart as a separate worksheet in the current workbook. If a name is not typed, Excel provides a default name of Chart 1, Chart 2, etc. The second choice, titled <u>As object</u> in, will place the chart as an object in the selected worksheet. Choose a location for the chart, and click the [Finish] button. The finished XY Scatter chart depicted in Figure 5-7 should appear.

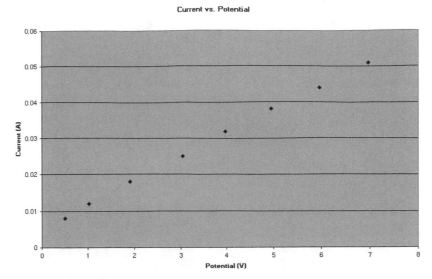

Figure 5.7.   The completed XY Scatter chart

5.1.2 Creating a Chart Using Shortcut Keys

If you frequently create charts of the same type, the **F11** shortcut key may be helpful. A chart that is created by using the **F11** key is formatted in the default chart type. The default chart type can be changed by choosing **Chart**, **Chart Type** from the Menu bar. The Chart Type dialog box depicted in Figure 5-8 will appear. Choose a chart type and subtype, and click on **Set as default chart**. Finally, choose **Cancel** instead of **OK** to exit the dialog box.

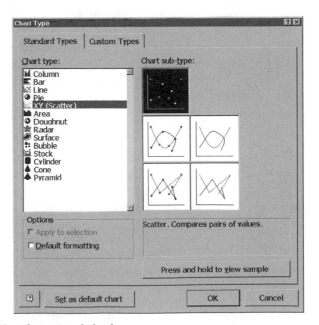

Figure 5.8.   The Chart Type dialog box

PRACTICE!

Practice creating a chart using the Chart Wizard by following the previously given steps.

Afterwards, practice creating a chart using the **F11** key. First, set the default chart type to <u>XY scatter</u>. Then select the data and headings from Figure 5-1. If you press the **F11** key, a chart similar to the one depicted in Figure 5-9 should appear. Note that the chart title and labels could use improvement.

Fortunately, chart elements can be easily modified. In Section 5.3, you will be shown how to modify and format charts.

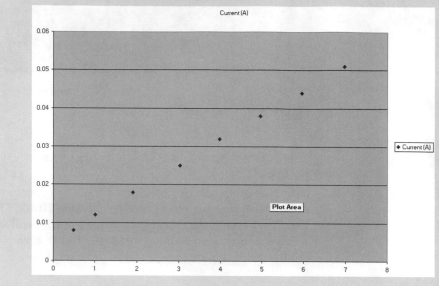

Figure 5.9. A chart created using the **F11** shortcut key

Excel makes several assumptions when automatically creating a chart. If the data do not follow Excel's conventions, then surprising results can sometimes occur. If the chart does not turn out as expected, the chart can always be reformatted as described in Section 5.3.

- Excel orients the chart so that the data for the X category are taken from the longest side of the selected region. In our example (from Figure 5-1), the longest side of the selected region is vertical.
- If the contents of the cells along the short side of the selected region contain text, that is used as labels for the data series in the legend. In our example, the label <u>Current (A)</u> is taken from the top cell of the short (horizontal) side of the selected region. If the cells contain numbers, the default data series names are used (Series 1, Series 2, etc.).
- If the contents of the cells along the long side of the selected region contain text, that text is used as X-category labels. If the cells contain numbers, then Excel assumes that the cells contain a data series.

5.1.3 Previewing and Printing Charts

To preview a chart before printing, either choose **File**, **Print Preview** from the Menu bar or choose the 🗋 button from the Standard toolbar. A chart that is embedded within a worksheet will print with the worksheet by default. If the chart is selected before choosing Print Preview, then the embedded chart can be printed separately. A chart that is formatted as a separate worksheet will, by default, be printed separately.

As an example, select the sheet containing the example XY Scatter chart in Figure 5-7, and choose 🔍 from the Standard toolbar. The Print Preview dialog box should appear. Select the **Margins** button, and the margin lines depicted in Figure 5-10 will appear. The margin lines can be dragged to resize and reshape the chart. The **Zoom** button can be used to focus on details of the chart.

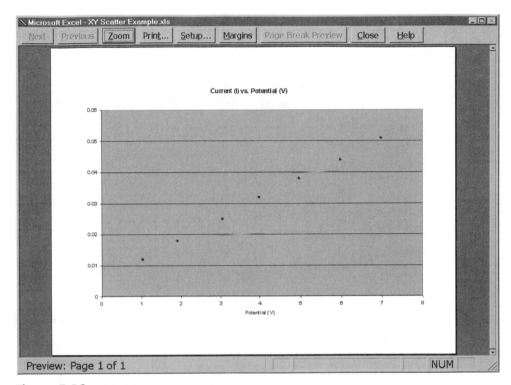

Figure 5.10. Print Preview options

Choose the **Setup** button to select more print formatting options. The Page Setup dialog box depicted in Figure 5-11 will appear. From the tabs on this dialog box, you may select portrait or landscape mode, select the chart size, select the printing quality, manually specify margins, and insert headers or footers.

5.2 ADDING AND EDITING CHART DATA

5.2.1 Adding Data Points

Data can be added or removed from a chart by activating the relevant chart and then choosing **Chart**, **Add Data** from the Menu bar. The Add Data dialog box depicted in Figure 5-12 will appear. As shown in the figure, we have added another data point in the experiment with Ohm's law.

After you have selected the new data range, the Paste Special dialog box depicted in Figure 5-13 will appear. Check the box labeled New point(s), since we want to add a

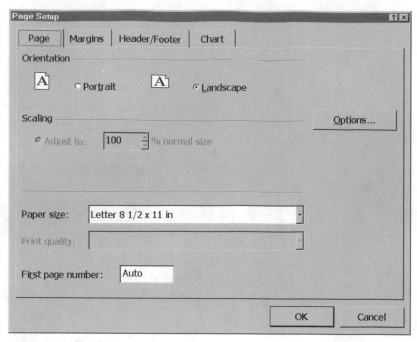

Figure 5.11. The Page Setup dialog box

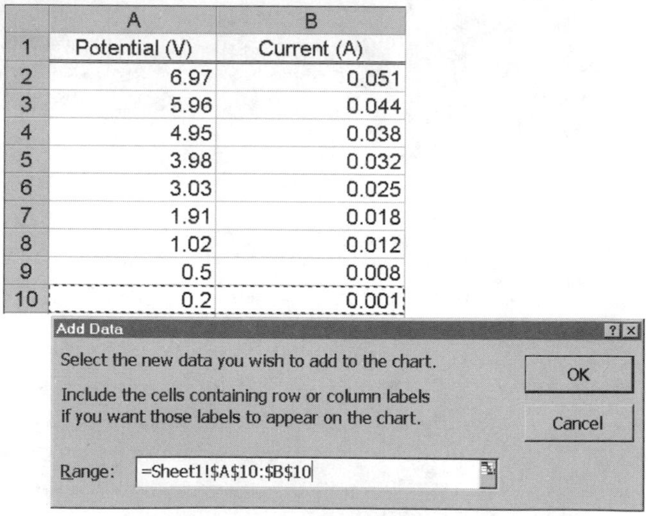

Figure 5.12. Adding data to a chart

new data point to the existing data series. Check the box labeled <u>Columns</u>, since in our example, the Y values (Current) are listed in columns. Check the box labeled <u>Categories (X Values) in First Column</u>, since our X values (Potential) are listed in the first column. Click **OK** and the new data point will be added to the chart.

Another method for adding (or deleting) data points is to modify the source data definition. Activate the chart by clicking once on it, and choose **Chart**, **Source Data** from the Menu bar. The Source Data dialog box depicted in Figure 5-14 will appear.

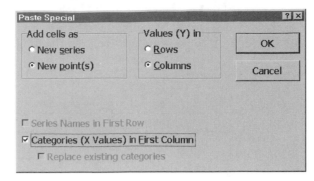

Figure 5.13.   The Paste Special dialog box

Choose the **Data Range** tab, and modify the contents of the box labeled Data range to add, modify, or delete source data points.

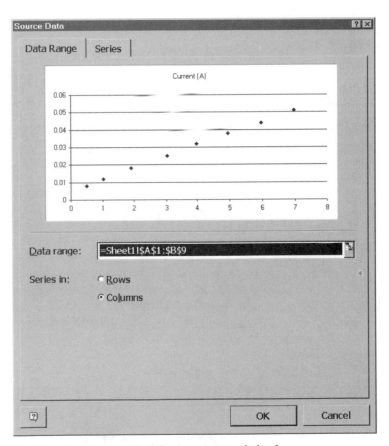

Figure 5.14.   The Data Range tab of the Source Data dialog box

5.2.2 Multiple Data Series

In the examples of charts so far, we have used a single *data series*. A data series is a collection of related data points that are to be represented as a unit. For example, the points in a data series are connected by a single line in a line chart. The data for a series would likely be represented by a separate row or column in the worksheet.

Table 5-1 shows data representing the flow rate for two tributaries of a river. The lowest one-day flow rate (in cubic feet per second) per year is shown. Create a worksheet that contains the data in Table 5-1, and use it for the rest of the examples in this chapter.

TABLE 5-1   Annual flow rate of two river branches

YEAR	EAST BRANCHFLOW (CFS)	WEST BRANCHFLOW (CFS)
87	221	222
88	354	315
89	200	175
90	373	400
91	248	204
92	323	325
93	216	188
94	195	202
95	266	254
96	182	176

Once you have created a worksheet containing the data and titles in Table 5-1, select the region and choose **Insert**, **Chart** from the Menu bar, or choose the 📊 button from the Standard toolbar. Follow the Chart Wizard instructions. Select **Line Chart** for a chart type, and select the first line chart subtype. The second Chart Wizard box will appear. Select the **Series** tab. The results should resemble Figure 5-15.

The Chart Wizard has created three data series, one for each column in the selected region of the worksheet. Since we would like the first column (Year) to be the X-axis data labels, remove the series titled Year from the Series box by selecting Year and clicking the **Remove** button. Add the range for the Year column to the box titled Category (X) axis labels. Complete the Chart Wizard. The resulting chart (without chart titles and axis titles) is depicted in Figure 5-16. Save this chart and use it as an example in the next section on chart formatting.

5.3 FORMATTING CHARTS

5.3.1 Creating Chart Objects

A chart consists of a number of elements called *chart objects*. Examples of chart objects are the chart legend and the chart title. Each object can be separately formatted and customized. It is usually faster to create the chart objects with the default formatting first. This will give you the general look of the chart. Then you can customize each object separately.

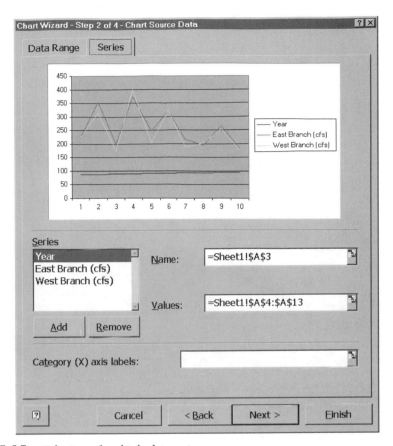

Figure 5.15. Selection of multiple data series

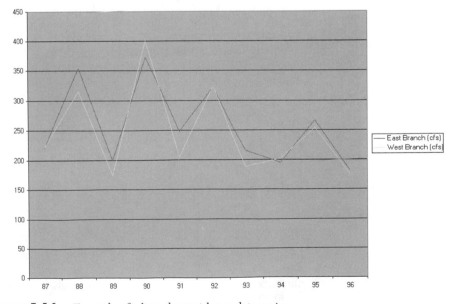

Figure 5.16. Example of a line chart with two data series

To create and enter data into chart objects, choose **Chart**, **Chart Options** from the Menu bar. The Chart Options dialog box depicted in Figure 5-17 will appear. The Chart Options dialog box contains seven tabs that are used, respectively, for entering and formatting the chart titles, axes, grid lines, legend, data labels, and data table. These are generally self-explanatory and will not be explained in detail here.

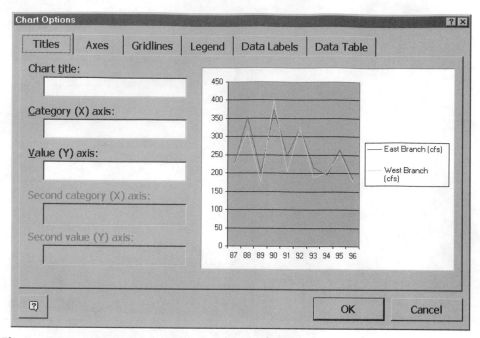

Figure 5.17. The Chart Options dialog box

5.3.2 Formatting Chart Objects

Any chart object can be formatted by selecting it with the left mouse button. When selected, an object is surrounded by a gray box. If the mouse cursor is positioned over the object, the name of the object is displayed. Figure 5-18 depicts the selected legend from Figure 5-16.

Figure 5.18. A selected chart object (Legend)

To format the selected object, click the right mouse button and choose **Format Legend Entry**. The Format Legend dialog box depicted in Figure 5-19 will appear. From the Format Legend dialog box, the border patterns, colors, font characteristics, and placement of the object can be customized. Another method for placing an object in the chart is to drag the object to the desired location.

This method of selecting an object with the mouse will work for formatting any chart object. Experiment by selecting, in turn, the X-axis, the Y-axis, and each of the

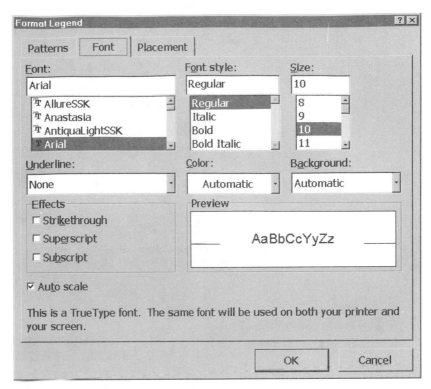

Figure 5.19. The Format Legend dialog box

data series. View the formatting dialog boxes for each object. The entire chart can be selected by clicking near the edge of the chart area. The plot area can be formatted by clicking in the rectangular plot area.

PROFESSIONAL SUCCESS: FORMATTING CHARTS

The appearance of a chart in a document is important. A chart can make a lasting visual impression that summarizes or exemplifies the main points of your presentation or document. The following formatting guidelines will help you create a professional-looking chart:

1. A chart title should contain a clear, concise description of the contents of the chart.

2. Create a label for each axis that contains, at a minimum, the name of the variable and the units of measurement that were used.

3. Create a label for each data series. The labels can be consolidated in a legend if each data series is represented by a distinct color or texture.

4. Scale graduations should be included for each axis. The graduation marks may take the form of grid lines or tick marks. The choice of marks can be manipulated somewhat by selecting the **Scale** tab from the Format Axis dialog box.

5. Ideally, graduations should follow the *1, 2, 5 rule*, which states that one should select the graduations so that the smallest division of the axis is a positive or negative integer power of 10 times 1, 2, or 5. For example, a scale graduation of 0.33 does not follow the rule.

PRACTICE!

Experiment with the chart in Figure 5-16.

- Add a chart title and titles for each axis.
- Format and move the legend.
- Change the plot background color.

After you have experimented with the chart in Figure 5-16, try to format your chart so that it resembles Figure 5-20 as closely as possible.

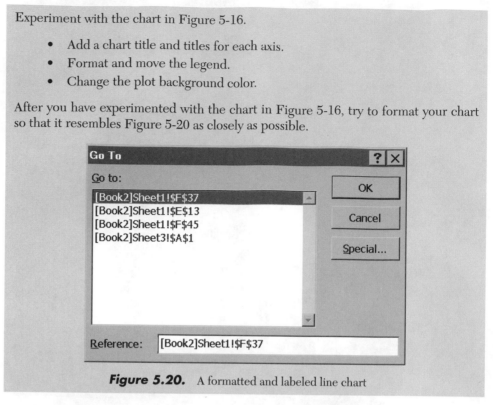

Figure 5.20. A formatted and labeled line chart

5.3.3 Changing Chart Types

The type of chart displayed can be changed after a chart has been created. The same data are used for the new chart type. Not all types of charts are appropriate for some data sets. For example, a pie chart is not appropriate for the data in Table 5-1. However, a bar chart *is* an acceptable chart type for the data in that table. To change the chart type for the chart in Figure 5-20, select the chart and choose **Chart**, **Chart Type** from the Menu bar. Select **Column** from the box labeled <u>Chart Type</u>, and select the first style from the box labeled <u>Chart sub-type</u>. The resulting chart should resemble Figure 5-21. Note also that the chart in your worksheet has been replaced. Note that the line chart is gone.

5.3.4 Inserting Text

The text for titles and axis labels can be added or modified by selecting and formatting chart objects. This is described in the previous section titled <u>Formatting Chart Objects</u>. At times, you may wish to add free-floating text to a chart. Free-floating text can be used to highlight or explain a specific data point. The functions on the Drawing toolbar may be used to create arrows or other free-floating shapes.

     As an example, we will add text and an arrow to the chart in Figure 5-20 to emphasize that 1989 was a low flow year for both river branches. To create the free-floating text:

1. Select any nontext object in the chart (e.g., the outside border).
2. Type the text and press the **Enter** key. In our example, we typed <u>1989 was a low flow year.</u>

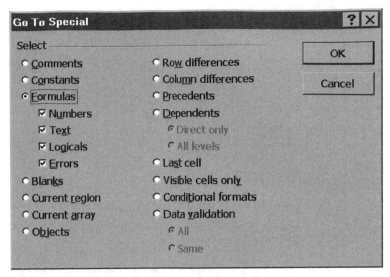

Figure 5.21. Example of columnar bar chart using data from Table 5-1.

The text should appear inside a small gray box. The font may be modified by choosing **Format, Text Box** from the Menu bar. If you are increasing the font size, the text box must be large enough to hold the new font. The box may be moved or resized with the mouse. The text can be edited directly on the chart simply by selecting the text box and then typing.

An arrow can be added by using the Drawing toolbar. If the Drawing toolbar does not appear on your screen, then choose **View, Toolbars** from the Menu bar and check the box titled Drawing. Practice creating a text box and arrow that resemble those of Figure 5-22.

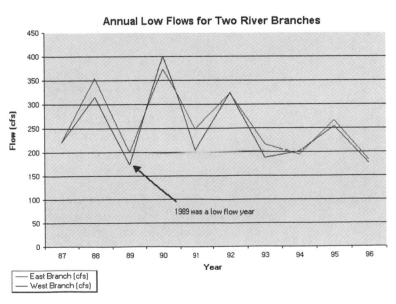

Figure 5.22. Chart with free-floating text and arrow

5.4 CHARTING FEATURES USEFUL TO ENGINEERS

Excel has many advanced features for formatting charts. Some of the features are particularly useful for engineering applications. These useful features include trend lines, error bars, what-if analysis, axis scaling, and secondary axes. The topics pertaining to data analysis (trend lines and what-if analysis) are covered in Chapter 6.

5.4.1 Scaling an Axis

The scale of an axis is used to delimit the range of the axis, as well as the intervals between axis markers called *ticks*. Large tick marks are specified in *major units*, and small tick marks are specified in *minor units*.

To change the axis formatting, click the left mouse button on the axis until the axis is highlighted, and then choose **Format, Selected Axis** from the Menu bar. The Format Axis dialog box depicted in Figure 5-23 will appear as. Select the **Scale** tab.

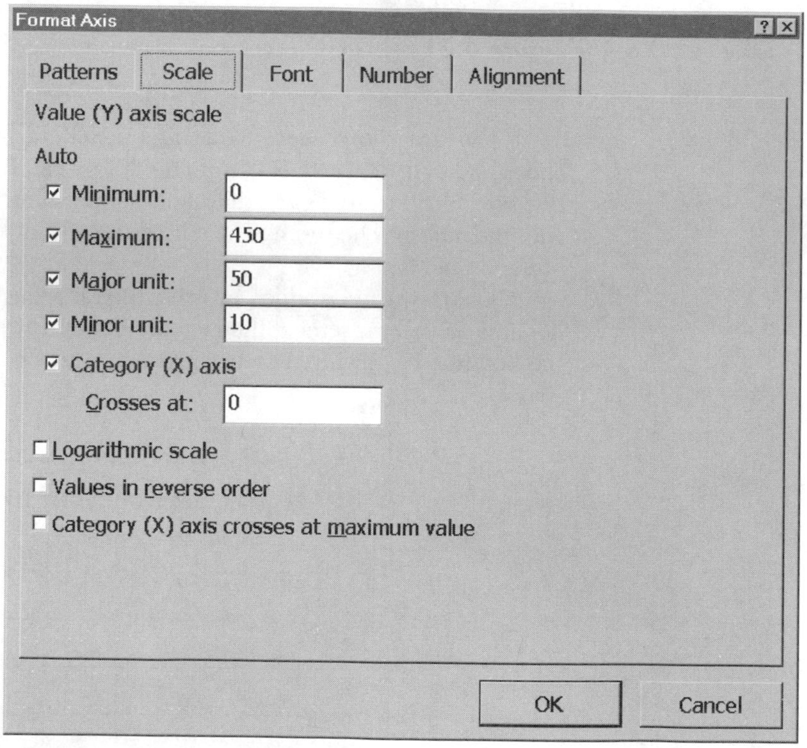

Figure 5.23. The Format Axis dialog box

When the boxes labeled **Minimum, Maximum, Major unit**, and **Minor unit** are checked, Excel will automatically choose appropriate values for the items. If you uncheck the boxes, then you can customize the items.

Of particular interest to engineers is the logarithmic scale. If you select this box, the values of the **Minimum, Maximum, Major unit**, and **Minor unit** boxes will be recalculated to be powers of 10. A logarithmic axis cannot contain values that are less than or equal to zero. Invalid values will produce an error message.

5.4.2 Error Bars

Error bars represent the range of statistical error in a data series. Error bars should not be used unless you understand their purpose. To add error bars to a data series, first select the series by clicking on it once with the left mouse button. Next hoose **Format**, and **Selected Data Series** from the Menu bar. Select the **Y Error Bars** tab. The Format Data Series dialog box depicted in Figure 5-24 will appear.

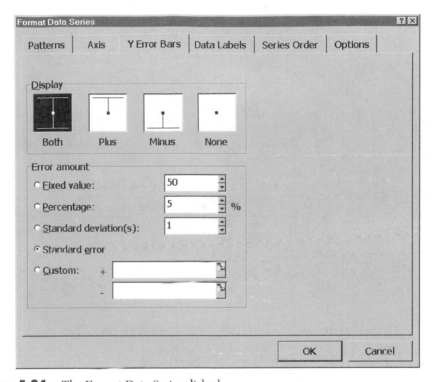

Figure 5.24. The Format Data Series dialog box

Choose a display type and an error amount. You can set a fixed or percentage amount if you have information about the error in the data series. Alternatively, you can let Excel calculate the standard deviation or standard error of the series for you. After you have created error bars, they will automatically be updated if the values in the data series change.

SUMMARY

In this chapter, the methods for creating and formatting charts were described. The chapter emphasize line charts and XY scatter plots, since these types of charts are frequently used in scientific and engineering applications. The available options for formatting chart elements were examined. Several topics applicable to engineering, scaling axes, and creating error bars were also presented.

KEY TERMS

1, 2, 5 rule   major units
chart objects   minor units
data series   Ohm's law
error bars   ticks

Problems

1. Generate data points for the function

$$y = 4\sin(x) - x^2$$

for $x = -5.0, -4.5, \ldots, 4.5, 5.0$. Chart the results using an XY scatter plot.

2. Add appropriate title and axis labels for the chart in Exercise 1. Try changing chart types. Experiment with line charts and 3D views.

3. A graph that uses logarithmic scales on both axes is called a log—log graph. This kind of graph is useful for plotting power equations, since they appear as straight lines. A power equation has the form

$$y = ax^b.$$

Table 5-2 presents data collected from an experiment that measured the resistance of a conductor for a number of sizes. The size (cross-sectional area) was measured in millimeters squared, and the resistance was measured in milliohms per meter. Create a scatter plot of these data.

TABLE 5-2   Resistance vs. area of a conductor

AREA A (MM2)	RESISTANCE (MILLIOHMS PER METER)
0.009	2000.0
0.021	1010.0
0.063	364.0
0.202	110.0
0.523	44.0
1.008	20.0
3.310	8.0
7.290	3.5
20.520	1.2

4. Modify both the X and Y axes of the chart created in the previous problem to use a logarithmic scale. What can you infer about the relationship between resistance and the size of a conductor in this experiment from viewing the resulting scatter plot?

6

Performing Data Analysis

6.1 TREND ANALYSIS

Trend analysis is the science of forecasting or predicting future elements of a data series on the basis of historical data. Trend analysis is used in many areas, such as financial forecasting, epidemiology, capacity planning, and criminology. Excel has the ability to calculate linear and exponential growth trends for data series. Excel can also calculate and display various trend lines for charts.

6.1.1 Trend Analysis with Data Series

A trend analysis can either extend or replace a series of data elements. The simplest method for extending a data series with a linear regression is to drag the fill handle past the end of the series. For example, Figure 6-1 shows the number of occurrences of a hypothetical disease for the years 1993 to 1996. Assuming a linear rate of increase in the disease, the number of occurrences can be estimated for 1997, 1998, and 1999.

To calculate the linear trend, select the known data (A4:D4), and drag the fill handle to the right so that the fill box covers cells (E4:G4). When you release the mouse, cells (E4:G4) will contain the data elements predicted by a linear regression of the original data. (See Figure 6-2.)

The Fill Series command can be used for somewhat more sophisticated trend analysis. To extend or replace a data series using the Fill Series command, select the region of data over which the analysis is to occur. This includes the cells with the original data and the new cells that are to hold the predicted data. Using the example in Figure 6-1, select cells (A4:G4). Choose **Edit**, **Fill**, and then **Series** from the Menu bar. The Series dialog box depicted in Figure 6-3 will appear.

OBJECTIVES

After reading this chapter, you should be able to:

- Calculate linear and exponential trends for data series
- Project trend lines onto charts
- Perform visual interactive analysis using Pivot Tables
- Undertake the iterative solution of equations using the Goal Seeker
- Access and use the Analysis ToolPak
- Solve optimization problems with the Solver feature

APPLICATION

Solving of constrained nonlinear optimization problems is an important task for engineers in the petroleum, chemical, defense, financial, agriculture, and process control industries. The Excel Solver may be the most widely used optimization software in the world. We will examine this tool in Section 6.5. The GRG2 code was developed by Leon Lasdon, University of Texas at Austin, and Allan Waren, Cleveland State University.

The following report was produced by Excel Solver for the unconstrained (except for nonnegativity) optimization problem with the objective function

$$y = 100\left(x_2 - x_1^2\right)^2$$

After reading this chapter, you will be able to apply the Solver to a variety of linear and non-linear optimization problems.

Microsoft Excel 8.0 Answer Report
Report Created: 6/14/98 4:22:33 PM

Target Cell (Min)

Name	Original Value	Final Value
y=	1	6.49784E-09

Adjustable Cells

Name	Original Value	Final Value
x_1	0	0.999977081
x_2	0	0.999961891

Constraints

Name	Cell Value	Formula	Status	Slack
x_1	0.999977081	$C$4>=0	Not Binding	0.9999771
x_2	0.999961891	$C$5>=0	Not Binding	0.9999619

	A	B	C	D	E	F	G
1	Occurrence of Disease X (in thousands)						
2							
3	1993	1994	1995	1996	(1997)	(1998)	(1999)
4	1.1	1.9	3.0	3.8			

Figure 6.1.   A sample set of disease data

	A	B	C	D	E	F	G
1	Occurrence of Disease X (in thousands)						
2							
3	1993	1994	1995	1996	(1997)	(1998)	(1999)
4	1.1	1.9	3.0	3.8	4.8	5.7	6.6

Figure 6.2.   Projected disease occurrences for 1997-99 using linear bestfit

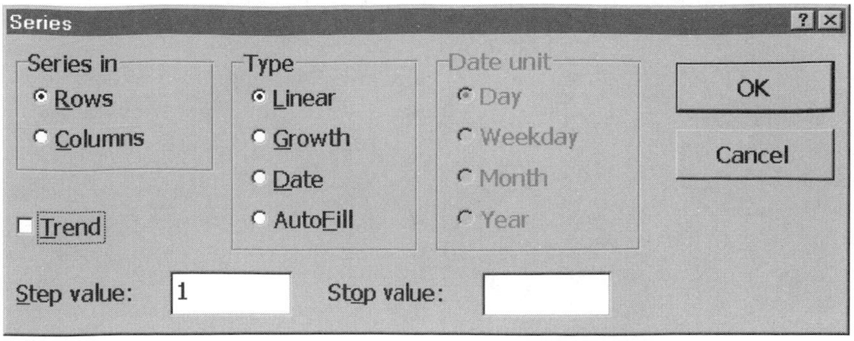

Figure 6.3. The Series dialog box

Note that a *step value* of one has been calculated by Excel. The step value can be modified manually to set the increment value for *x* in the linear equation $y = mx + b$. A *stop value* may be entered if you want to set an upper limit to the trend.

To extend the known values with a linear regression and leave the original values unchanged, check the box labeled <u>AutoFill</u> and uncheck the box labeled <u>Trend</u>. The results from the example are depicted in Row 6 (labeled Linear Extension) in Figure 6-4. Note that these results are identical to those obtained by dragging the fill handle in Figure 6-2.

To calculate a linear trend line and replace the original data values with best-fit data, check the box labeled <u>Linear</u> and the box labeled <u>Trend</u>. The trend line is no longer forced to pass through any of the original data points. The results are depicted in Row 8 (labeled Linear Replacement) in Figure 6-4.

A trend line using exponential growth can be calculated by checking the box labeled <u>Growth</u> and the box labeled <u>Trend</u>. The original data values are replaced, and the trend line is not forced pass through any of the original data points. The results are depicted in Row 10 (labeled Exponential Replacement) in Figure 6-4.

	A	B	C	D	E	F	G	H
1		Occurrence of disease X (in thousands of cases)						
2								
3		1993	1994	1995	1996	1997	1998	1999
4	Original Known Values	1.1	1.9	3.0	3.8			
5								
6	Linear Extension	1.1	1.9	3.0	3.8	4.8	5.7	6.6
7								
8	Linear Replacement	1.1	2.0	2.9	3.8	4.8	5.7	6.6
9								
10	Exponential Replacement	1.2	1.8	2.7	4.1	6.3	9.5	14.5
11								
12	TREND() function	1.1	2.0	2.9	3.8	4.8	5.7	6.6

Figure 6.4. Examples of trend analysis options

6.1.2 Trend Analysis Functions

Excel provides two trend analysis functions, one for calculating linear trends and another for calculating exponential trends. These are useful if the known dependent data may change and the trend line must be recalculated frequently. The linear trend function TREND uses the same least squares method that was used in the previous section. The syntax for TREND is

TREND *(known_y's, known_x's, new_x's, const)*

The arguments are described as follows.

Known_y's.   The known y values are the known dependent values in the linear equation $y = mx + b$. In Figure 6.4, the known y values are 1.1, 1.9, 3.0, and 3.8 in the range (B4:E4).

Known_x's.   The known x values are the values of the independent variable for which the y values are known. In Figure 6.4, these are the values 1993, 1994, 1995, and 1996 in the range (B3:E3). If the known x values are omitted, then the argument is assumed to be {1,2,3,4,…}.

New_x's.   The new x values are the values of the independent variable for which you want new y values to be calculated. If you want the predictions for years 1997 to 1999, select the range (F3:H3). If you want to calculate the linear trend for the whole time span (1993–1999), select the range (B3:H3).

Const.   If the *const* argument is set to FALSE, then b is set to zero, so the equation describing the relationship between y and x becomes $y = mx$. If the const argument is set to TRUE or omitted, then b is computed.

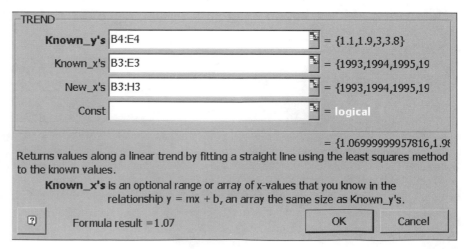

Figure 6.5.   The TREND function dialog box

To use the TREND function with our example, select a region in which to place the results. The region should be the same size as the number of *new_x's* to be calculated. Choose the **Edit Formula** button **=**, and select the TREND function from the function list. The TREND function dialog box depicted in Figure 6-5 will appear. Select or type the ranges described. Since, in this example, the arguments to TREND are arrays rather than single values, press **Shift** + **Ctrl** + **Enter**. The results are depicted in Row 12 (labeled <u>TREND() function</u>) of Figure 6-4. Since we selected the entire range of years for our *new_x's*, the results are identical to the results for linear replacement. The *known_x's* values can now be modified, and the trend results will be immediately recalculated.

There is a corresponding Excel function for predicting an exponential trend named GROWTH. The arguments to the GROWTH function are similar to the arguments for TREND.

6.1.3 Trend Analysis for Charts

Excel will calculate and display trend lines on a chart. Five types of regression lines can be added, or a moving average can be calculated. Each type of trend line is described in Table 6-1. Trendlines cannot be added to all types of charts. For example, trendlines cannot be added to data series in pie charts, 3D charts, stacked charts, or doughnut charts. Trendlines can be added to bar charts, XY scatter plots, and line charts. If a trendline is added to a chart and the chart type is subsequently changed to one of the exempted types, the trendline is lost.

TABLE 6-1   Formulas used to calculate chart trend lines

TYPE OF TREND LINE	FORMULA
Linear	The least squares fit is calculated using $y = mx + b$ (m is the slope and b is a constant).
Logarithmic	Calculates least squares using $y = c \cdot \ln(x) + b$ (c and b are constants).
Polynomial	Calculates least squares for a line using $$y = b + c_1 x + c_2 x^2 + \dots + c_n x^n$$ ($b, c_1, c_2, \dots, c_n$ are constants, the order can be set in the Add Trend line dialog box, and the maximum order is 6).
Exponential	Calculates least squares using $y = ce^{bx}$ (c and b are constants).
Power	Calculates least squares using $y = cx^b$ (c and b are constants).
Moving Average	Each data point in a moving average is the average of a specified number of previous data points. The series is calculated using $$F_{(t+1)} = \frac{1}{N} \sum_1^N A_{t-j+1}$$ (N is the number of previous periods to average, A_t is the value at time t, and F_t is the forecasted value at time t).

To create a trend line, first create a chart of an acceptable type. As an sample, create an XY chart using the exponential growth data in Row 10 of Figure 6-4. Choose **Chart**, **Add Trendline** from the Menu bar. The Add Trendline dialog box depicted in Figure 6-6 will appear. Choose the **Type** tab. Since we know that the data represents exponential growth, first choose **Linear** from the section labeled <u>Trend/Regression type</u> section. A linear trend line will not fit the data and will show maximum contrast for the example. Then choose **Series1** from the list labeled <u>Based on Series</u>.

Now choose the **Options** tab on the Add Trendline dialog box, as depicted in Figure 6-7. The first section of the Options tab gives the option of naming the trend line. The second section can be set to predict data points prior to or after the input data. The third section can be used to set the *y*-intercept to a constant value. If the box labeled <u>Set intercept</u> is not checked, then the intercept is calculated. Check the box labeled <u>Display equation on chart</u>, and click **OK**. The resulting chart and trend line should resemble Figure 6-8. If you add a trend line of exponential type, the data and line will match exactly.

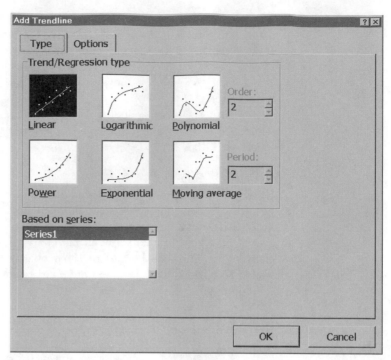

Figure 6.6.   The Type tab on the Add Trendline dialog box

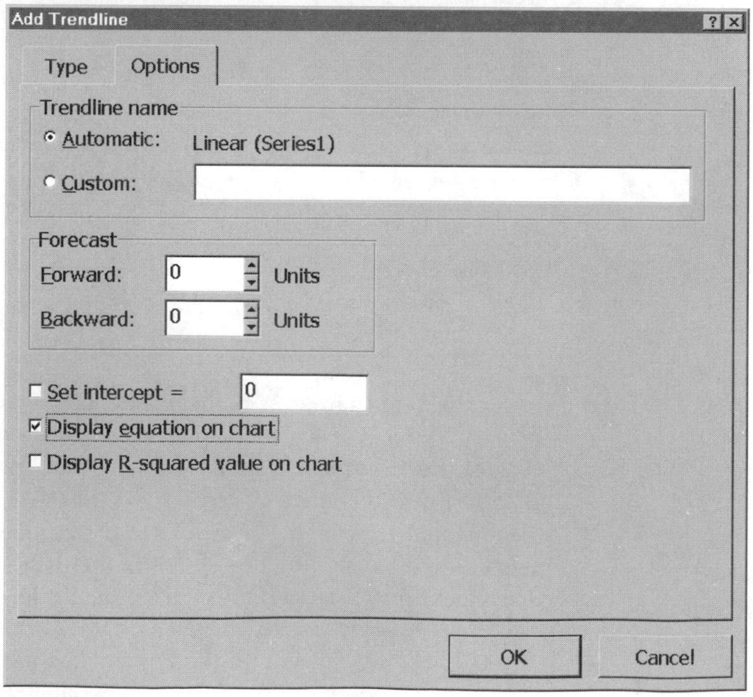

Figure 6.7.   The Options tab on the Add Trendline dialog box

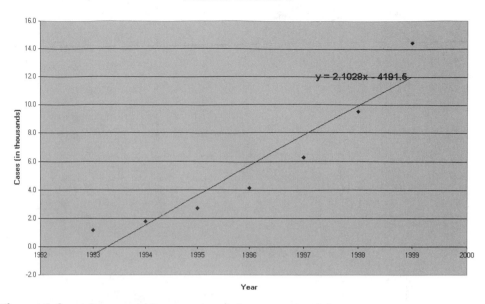

Figure 6.8. A linear trend line superimposed on exponential data

6.2 PIVOT TABLES FOR INTERACTIVE ANALYSIS

A *pivot table* is a table that summarizes data and that can be interactively manipulated. This frequently involves rotating or pivoting rows and columns–hence the name pivot table. The advantage of using a pivot table is the ease with which the data can be viewed from different perspectives.

A pivot table is a type of report. The data presented in the table cannot be manipulated directly. The underlying list can be modified, and the pivot table can be refreshed to display any resulting changes.

A pivot table is created from a list or database. The list can be an Excel worksheet, or it can be imported from an external source, which can be a database application, such as Access, dBase, or Oracle. The methods for obtaining external data from a database query are explained in Chapter 7. The external source can also be the World Wide Web. See Chapter 9 for an example for obtaining data using a Web query. The data for the example used in this chapter were obtained by using the Web query for Dow Jones Stocks by PC Quotes, Inc.

Before creating a pivot table, create or obtain a list of data to analyze. If you would like to practice whith the following example, create a worksheet that resembles Figure 6-9, or read Chapter 9 and obtain the data using a Web query.

To create a pivot table, use the PivotTable Wizard by choosing **Data**, **Pivot Table Report** from the Menu bar. Step 1 of the PivotTable wizard asks for the source of the data. Choose the box labeled Microsoft Excel list or database.

A portion of the stock quote data that we downloaded over the Web is depicted in Figure 6-10. In the following example, we are going to summarize the number of stocks in each category and then compute the average net change for each category.

Step 2 of the PivotTable Wizard asks for a range of data to summarize. Select the range of the sample worksheet that you created, or if you used a Web query, select the resulting worksheet.

	A	B
1	Category	Net Change
2	TRANSPORTATION	-1 3/4
3	TRANSPORTATION	-1 1/4
4	INDUSTRIALS	-1
5	INDUSTRIALS	- 5/8
6	TRANSPORTATION	- 9/16
7	INDUSTRIALS	- 3/8
8	INDUSTRIALS	- 3/8
9	UTILITIES	- 3/16
10	UTILITIES	- 3/16
11	UTILITIES	1/8
12	UTILITIES	1/4
13	INDUSTRIALS	5/8
14	INDUSTRIALS	2 1/2
15	TRANSPORTATION	3 7/16
16	INDUSTRIALS	6 5/8

Figure 6.9. Sample data for the pivot table example

	A	B	C	D	E	F	G
1	Company Name & Symbol	Category	Last Price	Net Change	Open	High	Low
4	AMERICAN EXPRESS COMPANY (AXP)	INDUSTRIALS	85 15/16	15/16	85 5/8	86 5/8	85 5
9	WALT DISNEY CO. (DIS)	INDUSTRIALS	93 9/16	-1 5/8	93 9/16	93 15/16	93
15	INTERNATIONAL BUSINESS MACHINES (IBM)	INDUSTRIALS	109 1/4	-1 1/8	110 5/8	111 5/16	108
18	J. P. MORGAN & CO., INC. (JPM)	INDUSTRIALS	122 3/4	2 1/2	120 1/4	124 1/4	119

Figure 6.10. Sample stock quotes obtained from a Web query

Step 3 asks you to construct the table by dragging *field buttons* into the report. The field buttons represent the field names of the database. In this example, drag the **Category** button into the box labeled <u>Row</u>. Then drag the **Category** button into the box labeled Data. The button will change its name to read **Count of Category** Finally, drag the **Net Change** button into the data area. The name is automatically changed to **Sum of Net Change**This is depicted in Figure 6-11.

Step 4 of the PivotTable Wizard asks you where to place the results. Choose the box labeled New Worksheet. Select the Finish button. The completed pivot table and the PivotTable toolbar depicted in Figure 6-12 will appear.

Now the benefits of using a PivotTable will become apparent. Recall that we want to compute the average of the stock's net change (not the sum). Select one of the cells labeled <u>Sum of Net Change</u>, and then choose the PivotTable Field button {missing} from the PivotTable toolbar. The PivotTable Field dialog box depicted in Figure 6-13 will appear.

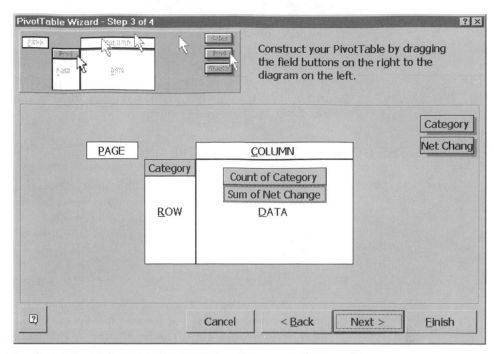

Figure 6.11. The PivotTable Wizard: Step 3

	A	B	C
1	Category	Data	Total
2	INDUSTRIALS	Count of Category	7
3		Sum of Net Change	7.375
4	TRANSPORTATION	Count of Category	4
5		Sum of Net Change	-0.125
6	UTILITIES	Count of Category	4
7		Sum of Net Change	0
8	Total Count of Category		15
9	Total Sum of Net Change		7.25
10			
11			
12			

Figure 6.12. Sample pivot table

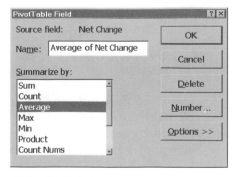

Figure 6.13. The Pivot Table Field dialog box

Select **Average** from the list labeled <u>Summarize by</u>, and then choose **OK**. The resulting pivot table will now show the average of the net stock changes for each category.

At any point, you may select the PivotTable Wizard button 🔳 from the PivotTable toolbar. This will immediately return you to Step 3 of the PivotTable Wizard. (See Figure 6-11.) From this box, you can drag the field buttons to new locations and quickly rearrange the PivotTable report.

6.3 USING THE GOAL SEEKER FOR WHAT-IF ANALYSIS

The *Goal Seeker* is used to find the input values of a formula when the results are known. The Goal Seeker takes an initial guess for an input value and uses iterative refinement to attempt to locate the real input value.

An example of the use of the Goal Seeker is the solution of a polynomial equation, for isntance,

$$f(x) = 3x^3 + 2x^2 + 4 = 0.$$

The equation $f(x)$ has a real solution that is approximately –1.3. If we can use the initial guess of -1.3 as a seed for the Goal Seeker, it will then attempt to converge on a more accurate value for x. The Goal Seeker does not always converge. With some functions, the initial guess must approximate the solution.

To see how the Goal Seeker works, create a worksheet that resembles Figure 6-14. Note the coding for $f(x)$ in the Formula box. The initial guess is $x = 1$. This results in $f(x) = 9$. Choose **Tools**, **Goal Seek** from the Menu bar. The Goal Seek dialog box depicted in Figure 6-14 will appear.

	A	B	C
1			
2			
3	f(x) = 3x³+2x²+4	f(x) =	9
4			
5		x =	1
6			
7		Goal Seek	
8		Set cell:	C3
9			
10		To value:	0
11		By changing cell:	C5
12			
13		OK	Cancel
14			

Figure 6.14.   Solving a polynomial equation with the Goal Seeker

Place the solution cell (C3), which contains the formula, in the box labeled <u>Set cell</u>. Place the desired solution (0) in the box labeled <u>To value</u>. Place the cell containing the initial guess (C5) in the box labeled <u>By changing cell</u>. Choose **OK**. The resulting value of $x = -1.373468173$ produces a solution $f(x) = 0.0000375$, which is close to zero.

6.4 USING THE ANALYSIS TOOLPAK

Another Excel add-in package includes a number of statistical and engineering tools. This package, called the *Analysis ToolPak*, can be used to shorten the time it usually takes to perform a complex analysis. To determine whether the ToolPak is installed on your system, choose **Tools** from the Menu bar. If the Data Analysis command does not appear on the Tools menu, then the Analysis ToolPak has not been installed or it has not been configured correctly. To install the Analysis ToolPak, run the Setup program on your Microsoft Office or Excel CD. After you have installed the ToolPak with the Setup program, load the add-in into Excel by choosing **Tools**, **Add-Ins** from the Menu bar. Check the box labeled <u>Analysis ToolPak</u> and click **OK**.

Once the ToolPak is successfully installed, choose **Tools**, **Data Analysis** from the Menu bar. The Data Analysis dialog box depicted in Figure 6-15 will appear. Browse through the list labeled <u>Analysis Tools</u>. Each tool requires a set of input parameters in a specific format, usually consist of an input range, an output range, and varying options. The results of the analysis are displayed in an output table. Additionally, some tools will generate a chart.

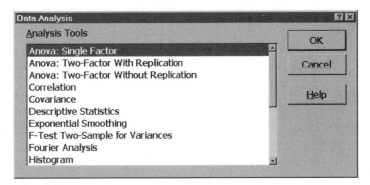

Figure 6.15. The Data Analysis dialog box

We will demonstrate the use of the Analysis ToolPak with the creation of a *histogram*, which is a graph of the frequency distribution of a set of data. The data are aggregated into classes, and the classes are graphed in a bar chart. The width of each bar represents the range of a class, and the height of each bar represents the frequency of data within a particular range. A class range is sometimes called a *bin*, since the histogram effectively places each data point into a bucket or bin.

An example of the use of a histogram is to graph the distribution of a set of student test scores for a course. A glance at the histogram can tell the instructor whether the test results are normally distributed or skewed. The two input ranges that are required for a histogram are a data set and a set of bin ranges.

The bin ranges are defined by listing the ascending boundary values. A data point is determined to be in a particular bin if the value of the data point is less than or equal to the bin number and greater than the previous bin number. You can choose to omit the bin range, in which case a set of evenly distributed bins between the data's minimum and maximum values is created.

Let us consider an example. Figure 6-16 shows a worksheet that contains the midterm grades of 20 students (B6:B25) and a set of boundary values for the classes or bins (D6:D13).

	A	B	C	D
5	Student#	Mid Term		Classes
6	1	45		30
7	2	89		40
8	3	90		50
9	4	67		60
10	5	88		70
11	6	93		80
12	7	32		90
13	8	85		100
14	9	68		
15	10	52		
16	11	77		
17	12	96		
18	13	54		
19	14	78		
20	15	83		
21	16	89		
22	17	79		
23	18	83		
24	19	72		
25	20	91		

Figure 6.16.   A group of student test scores and bin ranges

To create a histogram using the Analysis ToolPak, select **Histogram** from the Data Analysis dialog box. The Histogram dialog box depicted in Figure 6-17 will appear. Select the input range containing the test scores (B6:B25) and the bin range containing the bin boundaries (D6:D13). The output can be directed to a specified range, to a new worksheet, or a new workbook.

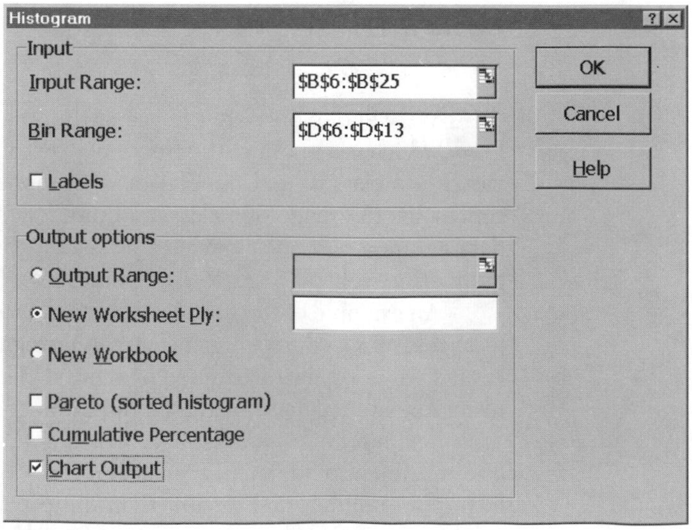

Figure 6.17.   The Histogram dialog box

Several optional features include sorting the histogram (<u>Pareto</u>), displaying a cumulative percentage (<u>Cumulative Percentage</u>), and charting the histogram (<u>Chart Output</u>). In the example, <u>Chart Output</u> is selected. The resulting frequency distribution table and chart are displayed in Figure 6-18.

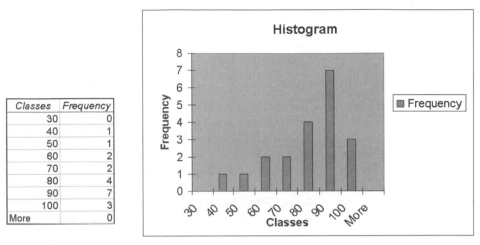

Classes	Frequency
30	0
40	1
50	1
60	2
70	2
80	4
90	7
100	3
More	0

Figure 6.18. Histogram of data in Figure 6-16

The Help button on the Histogram dialog box displays detailed information about the input parameters and options for the Histogram tool. The Help button can be used to access unique help for each of the Analysis Tools.

6.5 USING SOLVER FOR OPTIMIZATION PROBLEMS

Many engineering problems have more than one solution. Engineers choose among a range of possible solutions by applying limits to the input parameters. The problem then becomes one of finding a minimum (or maximum) solution, given the limitations. For example, you may want to minimize the cost of production of widgets, given the limitations of staff hours, availability of raw materials, power consumption, etc.

The equation that is to be maximized (or minimized) is called the objective function. The limitations to the input parameters of the *objective function* are called *constraints*. Problems of this type, such as finding a minimum or maximum given multiple constraints, are called *optimization* problems.

6.5.1 Introduction to Microsoft Excel Solver

Microsoft Excel provides a tool for solving optimization problems. Called Solver, this tool uses an implementation of the generalized reduced gradient algorithm called GRG2. The GRG2 code was developed by Leon Lasdon, of the University of Texas at Austin, and Allan Waren, of Cleveland State University. John Watson and Dan Fylstra, of Frontline Systems, Inc., implemented the methods used for linear and integer problems. Linear problems are solved using the simplex method with bounds on the variables, and integer problems are solved using the branch-and-bound method.

Solver must be installed as an add-in. To see if Solver is installed on your system, choose **Tools** from the Menu bar and look for the **Solver** menu item. If it is not present, then choose **Tools**, **Add-Ins** from the Menu bar. Choose **Solver** from the Add-Ins dialog box and click **OK**. If Solver does not appear on the list, then it should be installed from the Excel 97 or Office 97 installation CD.

6.5.2 Setting Up an Optimization Problem in Excel

The most difficult part of solving an optimization problem is setting up the objective function and identifying the constraints. An objective function takes the form

$$y = f(x_1, x_2, ..., x_n).$$

The independent variables are limited by m constraints, which take the form

$$c_i.(x_1, x_2, ..., x_n) = 0 \text{ for } i = 1, 2, ..., m.$$

The constraints may also be expressed as inequalities. Excel Solver can handle both linear and nonlinear constraints. However, nonlinear constraints must be continuous functions. Although the constraints are expressed as functions, they are always evaluated within a range of precision, called the *tolerance*. A constraint such as $x_1 < 0$ may be evaluated as TRUE when $x_1 = 0.0000003$ if the tolerance is large enough.

Nonlinear optimization problems may have multiple minima or maxima. In a minimization problem, all solutions except the absolute minimum are called *local minima*. The solution that is chosen by Solver is dependent on the initial starting point, which should be as close to the real solution as possible. These are problems with optimization in general, not with Excel Solver. For the rest of the examples in this section, it is assumed that you have some familiarity with optimization.

6.5.3 Linear Optimization

Suppose that you wish to maximize the profit for producing widgets, which come in two models: economy and deluxe. The economy model sells for $49.00 and the deluxe model sells for $79.00. The cost of production is determined primarily by labor costs, which are $14.00 per hour. The union limits the workers to a total of 2,000 hours per month. The economy widget can be built in 3 people-hours, and the deluxe widget can be built in 4 people-hours. The management believes that it can sell up to 600 deluxe widgets per month and up to 1200 economy widgets. Since you have a limited workforce, the main variable under your control is the ability to balance the number of economy vs. deluxe units that are built. Your job is to determine how many economy widgets and how many deluxe widgets should be built to maximize the company's profit.

The independent variables are

w_1 = the number of economy widgets produced each month

and

w_2 = the number of deluxe widgets produced each month.

The target that you wish to maximize is the profit p, which is described mathematically by the objective function

$$p = (49 - (3 \cdot 14))w_1 + (79 - (4 \cdot 14))w_2 = 7w_1 + 23w_2.$$

The constraints can be expressed mathematically as limitations on w_1 and w_2. The maximum number of widgets to be produced is limited by both sales and the availability of labor. The sales limitations can be expressed as

$$w_1 \leq 1200 \text{ widgets}$$

and

$w_2 = 600$ widgets.

The limitation imposed by the availability of labor can be expressed as

$3w_1 + 4w_2 \leq 2000$.

Finally, the general constraint of nonnegativity is imposed on w_1 and w_2, since you cannot produce a negative number of widgets. So,

$w_1, w_2 \geq 0$.

Figure 6-19 shows how to set up the widget problem in a worksheet. Cells D5 and D6 have been named w_1 and w_2, respectively, to make the formulas more readable. Note the underlines in the names. The names w1 and w2 cannot be used, since these are reserved for cell identifiers. The formulas are displayed in the cells for readability. To display formulas instead of their results, choose **Tools**, **Options**, **View** from the Menu bar. Check the box labeled Formulas.

	A	B	C	D	E
1		Widget Profit Optimization Worksheet			
2					
3					
4		Independent variables			
5	Ecomony widgets/month w1–			0	
6	Deluxe widgets/month w2=			0	
7					
8		Objective function			
9		Profit(p)	=7*w_1 + 23*w_2		
10					
11					
12		Constraints			
13		Labor constraint	=3*w_1 + 4*w_2		

Figure 6.19. Worksheet for linear optimization example

Choose **Tools**, **Solver** from the Menu bar. The Solver Parameters dialog box depicted in Figure 6-20 will appear. The box labeled Set Target Cell should contain the cell holding the objective function (the profit function). Check the box labeled Max, since you want to maximize profit. The box labeled By Changing Cells should contain the input parameters w_1 and w_2. Add each of the constraints as depicted in Figure 6-20, and choose **Options**.

The Solver Options dialog box depicted in Figure 6-21 will appear. Accept most of the options in the box. These should not be changed unless you understand the GRG2 and simplex methods used to implement solver. Since the problem you are working on is a linear one, check the box labeled Assume Linear Model, and then choose **OK** to return to the Solver Parameters dialog box. Choose **Solve** to begin the computation.

The Solver Results dialog box depicted in Figure 6-22 will appear. It should state that Solver has found a solution. There are three types of reports available from Solver. Select all three: **Answer**, **Sensitivity**, and **Limits**. Check the box labeled Keep Solver Solution and then press OK.

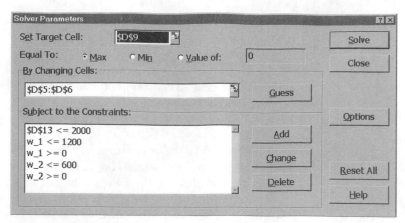

Figure 6.20.   The Solver Parameters dialog box

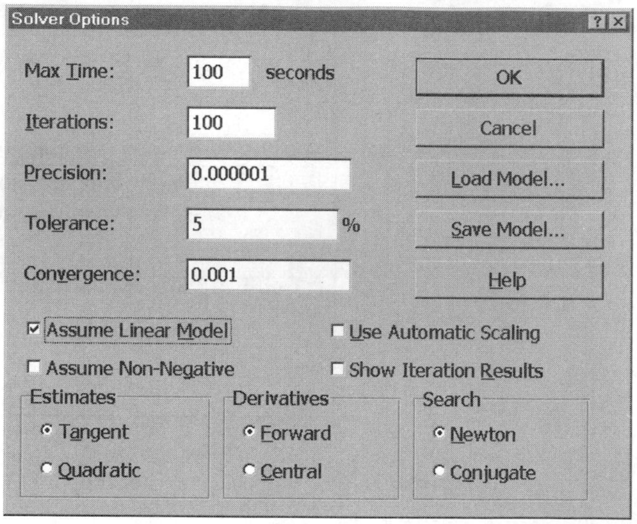

Figure 6.21.   The Solver Options dialog box

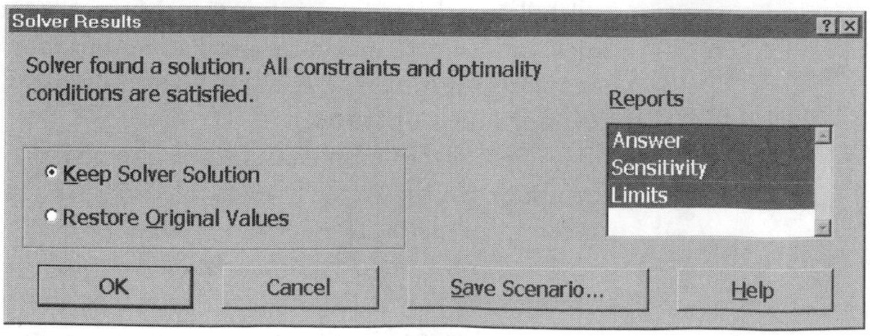

Figure 6.22.   The Solver Results dialog box

The worksheet that you created will now contain modified values for the input parameters and objective function. In addition, three new worksheets will be created, for the Answer, Sensitivity, and Limits Reports.

The Answer report summarizes the initial and final values of the input parameters and the optimized variable. The Sensitivity report gives information about marginal effects of making small changes in the constraints. Sometimes, a small change in a constraint can make a large difference in the output. For nonlinear models, the constraints are called Lagrange multipliers. For linear models, they are sometimes called dual values or shadow prices. The Limits report shows the effect on the solution as each input parameter is set to its minimum and maximum limit.

The resulting spreadsheet from our example is shown in Figure 6-23. A brief look at this worksheet shows that the maximum profit of $11,500 per month is achieved by producing only deluxe widgets. Five hundred deluxe widgets can be produced per month, but the company can sell 600 per month; thus, a limiting constraint is the available labor pool.

	A	B	C	D	E
1	Widget Profit Optimization Worksheet				
2					
3					
4	Independent variables				
5	Ecomony widgets/month w1=			0	
6	Deluxe widgets/month w2=			500	
7					
8	Objective function				
9	Profit(p)			11500	
10					
11					
12	Constraints				
13	Labor constraint			2000	

Figure 6.23. Results of linear optimization

As a manager, you can easily modify the constraints and rerun Solver to see the effect. You could rapidly view the effect on profit of hiring more laborers, modifying prices, or adjusting the widget mix.

6.5.4 NonLinear Optimization

As an example of nonlinear optimization, we will use an optimization problem to which the solution is obvious. This will familiarize you with the process of setting up a non-linear optimization problem, and it will convince you that the results are correct.

The objective function that we wish to minimize is

$$y = 100(x_2 - x_1^2)^2 + (1 - x_1)^2,$$

with the nonnegativity constraints

$$x_1 \geq 0$$

and

$$x_2 \geq 0.$$

Since the terms $(x_2 - x_1^2)^2$ and $(1 - x_1)^2$ must be positive for real numbers x_1 and x_2, the minimum y is zero, with $x_1 = 1$ and $x_2 = 1$. The worksheet for this example is shown in Figure 6-24.

	A	B	C	D	E
1	Example of non-linear optimization				
2					
3	Input Parameters				
4	x1=	0			
5	x2=	0			
6					
7	Objective Function				
8	y=	=100*(x_2 - x_1^2)^2 + (1-x_1)^2			

Figure 6.24.   Worksheet for nonlinear optimization example

Follow the steps described for linear optimization with the following changes:

- Check the box labeled <u>Min</u>.
- Set the constraints x_1 >= 0 and x_2 >= 0.
- Do *not* check the box labeled <u>Assume Linear Model</u>.

The following results produced by Solver are good approximations of the true minimum:

$x_1 = 0.999977$.

$x_2 = 0.999962$.

$y = 0.6.49784E -09$.

There are other local minima for this objective function. Try setting the initial parameters to $x_1 = 3$ and $x_2 = 5$. The results produced by Solver show that the algorithm gets stuck in a local minimum:

$x_1 = 1.639202$.

$x_2 = 2.668455$.

$y = 0.44290084$.

SUMMARY

This chapter introduced many of the powerful data analysis tools offered by Excel. These tools which include a large number of statistical and engineering functions in the Analysis ToolPak, a method for solving equations using the Goal Seeker, and methods for solving optimization problems using the Solver feature, can be used to slove problems in your engineering courses.

KEY TERMS

Analysis ToolPak
bin
field button
Goal Seeker
histogram

pivot table
step value
stop value
trend analysis

Problems

1. Generate the values of $f(x) = 2.5 \ln(x) + 1.34$ for $x = 1, 2, \ldots, 10$. Chart the results using an XY scatter plot. Create a logarithmic line for the data series. How well does the line match the data? Choose the options box on the Format Trendline dialog box, forecast forward 4 units, and display the equation on the chart. How closely does the equation match the original function?

2. Generate data points for the function
    $$y = 4\sin(x) - x^2.$$
    for $x = -5.0, -4.5, \ldots, 4.5, 5.0$. Chart the results using a line chart. Add a trend line to the chart. Display the equation on the chart. Does a fifth-order polynomial closely match the data?

3. Compute the values of
    $$f(x) = x^3 - 12x - 9$$
    for $x = -4.0, -3.5, -3.0, \ldots, 3.5, 4.0$. Chart the results using an XY scatter plot. Note that in this range $f(x)$ crosses the X axis three times, near $x = -3.0, -0.8$, and 3.8. Use the Goal Seeker to find more accurate solutions for
    $$x^3 - 12x - 9 = 0.$$

4. Use the method described in Problem 2 to find the solution of
    $$f(x) = x^3 + \sin(x / 2) + 2x - 4 = 0 \text{ for } -1.0 \le x \le 2.0.$$

5. Use the Descriptive Statistics selection from the Analysis ToolPak to find the mean and standard deviation for the student data in Figure 6-16. How many midterm grades lie more than two standard deviations from the mean?

7

Database Management within Excel

7.1 INTRODUCTION

Microsoft Excel implements a rudimentary database management system by treating lists in a worksheet as database records. This approach is helpful for organizing, sorting, and searching through worksheets that contain many related items. You can import complete databases from external *database management systems (DBMSs)*, such as Microsoft Access, Oracle, dBase, Microsoft FoxPro, and text files. You can create structured queries using Microsoft Query that will retrieve selected information from external sources.

If you require a relational DBMS, you are encouraged to use another, more complete software application, such as Microsoft Access. However, the database functions within Excel are adequate for many problems. An example of one way that an engineer might use this functionality is to import experimental data that have been stored in a relational DBMS to perform analysis on the data using Excel's built-in functions.

7.2 CREATING DATABASES

7.2.1 Database Terminology

A *database* within Excel is sometimes called a *list*. The two terms will be treated synonymously in this book. A *database* can be thought of as an electronic file cabinet that contains a number of folders. Each folder contains similar information about different objects. For example, each folder might contain the information about a student at a college of engineering. The database is the collection of all student folders. The data in each folder is organized in a similar fashion. For

OBJECTIVES

After reading this chapter, you should be able to:

- Create a database within Excel
- Enter data into a database
- Sort a database on one or more keys
- Use filters to search or collapse databases
- Use Microsoft Query to access external databases

APPLICATION

The amount of scientific information that is being accumulated is overwhelming. Scientific knowledge is being stored and catalogued in scientific databases at a phenomenal rate. New methods for searching databases and extracting relevant information are being devised.

One example of a large and fruitful endeavor to catalog and disseminate knowledge is the U.S. Human Genome Project. This effort, begun in 1990, is a 15-year cooperative program to identify all 80,000 genes in human DNA and to determine the sequences of the 3 billion chemical bases that make up human DNA.

Microsoft Query is an application accessible through Excel that can be used to query remote data from a variety of database formats. MS Query *Structured Query Language (SQL)* statements can be used to extract remote data and copy the results directly into an Excel worksheet.

In this chapter, you will learn how to create and manipulate databases within Excel. You will also learn how to acquire data from other database formats. In Chapter 9, you will learn how to query remote sites over the World Wide Web.

Image Created by: Dr. Edward H. Egelman, Univ. of Minnesota, Dept. of Cell Biology

example, each folder includes a student's first name, last name, social security number, address, department, class, etc.

In database terminology, each folder is called a *record*. Each data item, such as the student's first name, is stored within a *field*. The title for each data item is called a *field name*.

Any region in an Excel worksheet can be defined to be a database. Excel represents each record as a separate row. Each cell within the row is a field. The heading for each column is the field name.

Figure 7-1 depicts a small student database. Rows 2 through 7 each represent a student record. Each record has five fields. The field names are the column headings (e.g., Last Name).

	A	B	C	D	E
1	**Last Name**	**First Name**	**SSN**	**Department**	**Class**
2	Clinton	Willie	243-65-7666	Electrical	Junior
3	Smith	Randolph	245-54-3223	Chemical	Senior
4	Simpson	Susie	268-22-5365	Electrical	Senior
5	Smith	Christine	287-56-4532	Civil	Junior
6	Washington	Frank	532-45-3343	Mechanical	Junior
7	Granger	Linda	576-43-5455	Chemical	Senior

Figure 7.1.   Example of a student database

7.2.2 Database Creation Tips

Most database management systems store records in one or more files. The file delimits the boundaries of the database. Excel, however, stores a database as a region in a worksheet.

Excel must have some way of knowing where the database begins and ends in the worksheet. There are two methods for associating a region with a database. One method is to leave a perimeter of blank cells around the database region. The other is to explicitly name the region. Because of the unique way that Excel delimits a database, the following tips are recommended:

- Maintain only one database per worksheet. This will speed up access to the sorting and filtering functions, and you will not need to name the database regions.
- Each column heading in the database must be unique. If there were two headings for Last Name, for example, a logical query, such as Find all records with last name equal to Smith, would not make sense.
- Create an empty column to the right of the database and an empty row at the bottom of the database. Excel uses the empty row and column to mark the edge of the database. An alternative method is to assign a name to the region of the database. A disadvantage of assigning a name is that the allocated region may have to be redefined when records are added or deleted.
- Do not use cells to the right of the database for other purposes. Filtered rows may inadvertently hide these cells.

7.2.3 Entering Data

Once the field names for the database have been created in the column headings, the data may be entered using several methods. One method of data entry is to directly enter data into a cell. A database field may have any legitimate Excel value, including numerical, date, text, or formula. For example, you might add a column to the database in Figure 7-1 that is titled Full Name. Instead of copying or retyping the first and last names of each student, the new field could concatenate First Name and Last Name using the formula:

`=CONCATENATE(B2," ",A2).`

A second method for entering data is to use a form. To access the Data Entry form, click on any data cell in the database. Choose **Data**, **Form** from the Menu bar. The Data Entry form depicted in Figure 7-2 will appear. The title of the form will be the same as the name of the current worksheet.

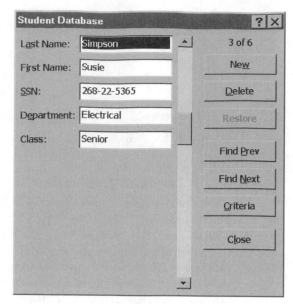

Figure 7.2. The Data Entry form

From the Data Entry form, a new record can be created by clicking the **New** button. From the form, you can scroll through the database, delete records, and modify existing records. The Data Entry form can also be used to filter data. This feature is explained in the next section.

PRACTICE!

Before proceeding, it would be helpful if you created the database depicted in Figure 7-1. This database will be used for the examples in Section 7.2.

Enter some of the data from the database using the Data Entry form. Enter some of the data by typing directly into the worksheet. Which method is less prone to typing errors?

7.3 SORTING, SEARCHING, AND FILTERING

The power of a database management system lies in its ability to search for information, rearrange data, and filter information.

7.3.1 Sorting

To sort a database, select any cell within the database and choose **Data**, **Sort** from the Menu bar. The Sort dialog box depicted in Figure 7-3 will appear. The field on which the sort is made is called the *sort key*. Excel allows you to sort on multiple keys. For example, choose **Last Name** in ascending order as the first key. Choose **First Name** in ascending order as the second key. Make sure that the box labeled Header Row is checked, and click **OK**. The result is an alphabetical listing of the student database.

Warning:

Be sure to select the entire database before sorting. If some columns are left out of the sort, the database may become scrambled. If you accidentally scramble the database, immediately choose **Edit**, **Undo Sort** from the Menu bar. The easiest way to select the

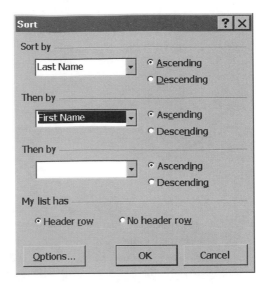

Figure 7.3. The Sort dialog box

entire database is to click on a single cell before performing any database operations. The entire database will be highlighted.

7.3.2 Searching and Filtering

Excel has several mechanisms for locating records that match specified criteria. For example, you may be interested in reviewing the students with chemical engineering majors. After you specify the criterion of Department to be Chemical, Excel displays only those records with the Department field equal to Chemical. The process of limiting the visible records based on one or more criteria is called *filtering*. There are three general methods for filtering a database in Excel. The easiest methods are the use of the Data Entry form and the AutoFilter function. The Advanced Filter function allows you to search using more logically sophisticated search criteria.

Filtering with the Data Entry Form.   To use the Data Entry form to filter and search a database, select a cell within the database and choose **Data**, **Form** from the Menu bar. Select the Criteria button on the Data Entry form. A blank record will appear. As an example, type **Smith** in the Last Name field. Now repeatedly click the **Find Next** and **Find Prev** buttons, and note that only students with a last name of Smith appear.

To clear the filter, choose **Criteria**, **Clear**, **Form** from the Data Entry form. Now if you click the **Find Next** and **Find Prev** buttons, the name of all of the students will appear.

Using the AutoFilter Function.   The AutoFilter function allows you to filter records while viewing the database as a worksheet. To turn on the AutoFilter function, select a cell within the database and choose **Data**, **Filter**, **AutoFilter** from the Menu bar. A small arrow will appear in the heading of each column. When you click on one of the arrows, a small drop-down menu will appear that contains the choices for that field. Figure 7-4 depicts the drop-down menu for the Department field.

	A	B	C	D	E
1	Last Name ▾	First Name ▾	SSN ▾	Department ▾	Class ▾
2	Clinton	Willie	243-65-7666	(All)	Junior
3	Smith	Randolph	245-54-3223	(Top 10...)	Senior
4	Simpson	Susie	268-22-5365	(Custom...) Chemical	Senior
5	Smith	Christine	287-56-4532	Civil	Junior
6	Washington	Frank	532-45-3343	Electrical	Junior
7	Granger	Linda	576-43-5455	Mechanical Chemical	Senior

Figure 7.4.   An sample AutoFilter menu

PRACTICE!

Practice using the AutoFilter function, and you will see how easy it is to define a filter and modify. First, Select the arrow in the Class field and choose **Senior**. The database will immediately hide all of the records except those of the seniors. Now select the arrow in the Department field and chose **Electrical**. The result is a filter that displays all of the students who are seniors *and* are also in electrical engineering. The two criteria are treated as if connected with a logical AND.

Note that the arrows for the Department and Class fields have changed color. This alerts you that a filter has been applied on these fields. To remove the filters, choose each colored arrow and select **(ALL)** from the drop-down menu. The arrow will return to its original color and all of the database records will reappear.

You can specify criteria that are more complex by selecting a field and choosing **(Custom)**. Try this by selecting the arrow in the **Department** field. Then choose **(Custom)**. The Custom AutoFilter dialog box depicted in Figure 7-5 will appear.

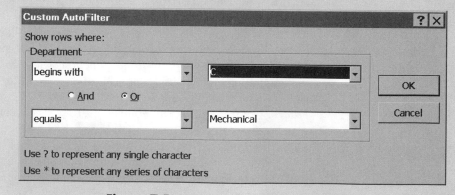

Figure 7.5.   The Custom AutoFilter dialog box

Experiment with the criteria. The selections in Figure 7-5 will create a filter that displays students whose department begins with a **C** or equals **Mechanical**. Try experimenting with the **?** and **°** wild cards. For example, try a custom filter using the criterion

```
SSN equals *45*.
```

Using the Advanced Filter Function.   The Advanced Filter function is useful if you have a complex set of criteria or you want to filter on calculated cells. Before creating a complex set of criteria, you must set up a criterion table. Do this by copying the field names to another location in the worksheet. Leave at least one blank row of cells between the table and the database.

A criterion table may have more than one active row. Every criterion within a single row must be met for a match to occur. This operation is equivalent to a logical AND. Figure 7-6 depicts a criterion table and a database. The first row of the table (Row 5) is equivalent to the English statement

```
Select the students whose GPA is greater than 3.5 AND who are
seniors.
```

The second row of the criterion table (Row 6) is equivalent to the English statement

```
Select the students whose GPA is greater than 3.1 AND who are
in the electrical engineering department.
```

The Advanced Filter accepts matches from either criterion. This operation is equivalent to a logical OR. The AND operator has precedence over the OR operator. The effect of both rows is equivalent to the English statement

```
Select the students

whose GPA is greater than 3.5 AND who are seniors,

OR

whose GPA is greater than 3.1 AND are in the electrical engi-
neering department.
```

	A	B	C	D	E
1	Criteria				
3	**Last Name**	**First Name**	**GPA**	**Department**	**Class**
5			>3.50		Senior
6			>3.10	Electrical	
7					
8	Database				
9	**Last Name**	**First Name**	**GPA**	**Department**	**Class**
10	Clinton	Willie	3.51	Electrical	Junior
11	Smith	Randolph	2.98	Chemical	Senior
12	Simpson	Susie	3.92	Electrical	Senior
13	Smith	Christine	3.65	Civil	Junior
14	Washington	Frank	3.41	Mechanical	Junior
15	Granger	Linda	2.78	Chemical	Senior
16					
17					
18	Query Results				
19	**Last Name**	**First Name**	**GPA**	**Department**	**Class**
20	Clinton	Willie	3.51	Electrical	Junior
21	Simpson	Susie	3.92	Electrical	Senior

Figure 7.6. A criterion table used with the Advanced Filter

7.4 EXTERNAL DATA RETRIEVAL USING MICROSOFT QUERY

A stand-alone program called Microsoft Query is capable of retrieving data from other types of databases into your worksheet. Although Microsoft Query is included with the Excel 97 package, it is not part of the default installation. If you have a version of Excel older than to Excel 97, Microsoft Query must be installed as an add-in. In addition, each type of database that can be accessed has a separate driver, called an Open Database Connectivity (ODBC) driver. Such drivers exist for Paradox, dBase, Access, etc. If Microsoft Query does not work, run the Excel 97 or Office 97 Setup program, and install Query and an ODBC driver for each type of database that you wish to access.

For the following examples, we have created a small dBase database with four fields and four records. This database is depicted in Figure 7-7.

Record#	NAME	SSN	PHONE	AGE
1	Dave Kuncicky	267863456	(850)487-6431	48
2	Steffie Grow	254534321	(850)421-3013	41
3	Helen Alice	564324543	(850)385-4960	34
4	Cliff Browning	342546543	(850)421-5465	46

Figure 7.7.   A sample dBase database

After you have installed Microsoft Query, select a cell and choose **Data**, **Get External Data**, **Create New Query** The Choose Data Source dialog box depicted in Figure 7-8 will appear. If the box labeled Use the Query Wizard is checked, remove the check. Select the database type from the list under the **Databases** tab. For this example, choose **dBase Files** from the list, and OK.

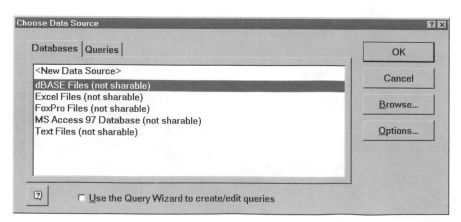

Figure 7.8.   The Choose Data Source dialog box

The Add Tables dialog box depicted in Figure 7-9 will appear. Locate the database from which you wish to import data, and click **Add**. The Microsoft Query program will then execute, and the database name and field names should appear in the MS Query window. Part of the MS Query window is depicted in Figure 7-10.

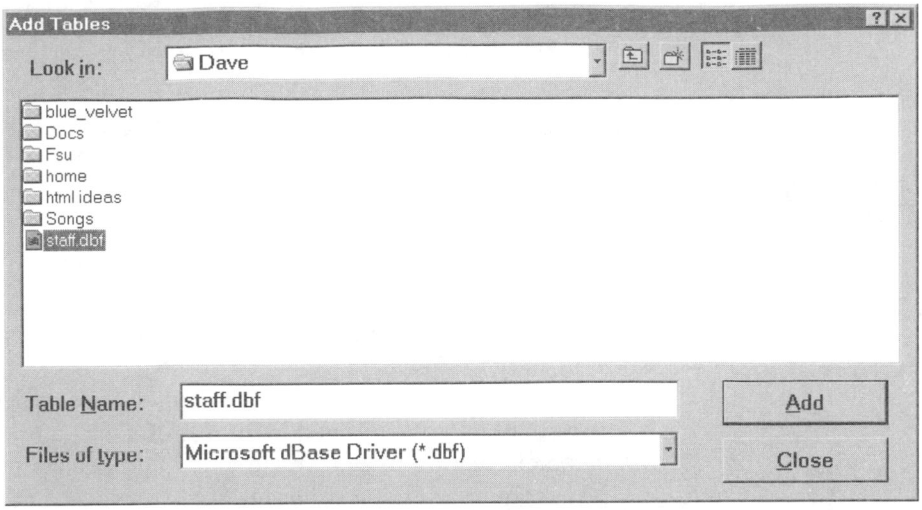

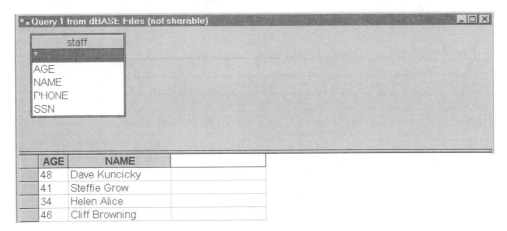

Figure 7.9.   The Add Tables dialog box

Figure 7.10.   Selecting fields from the external database

Note that the four fields in the dBase file (depicted in Figure 7-7) appear in the top pane of the MS Query window. You can select the fields of interest by clicking on the field names. The result will appear at the bottom of the window. In the example, the fields AGE and NAME were selected. (See Figure 7-10.)

Once the fields have been selected, the database may be filtered to limit the records that are displayed. If the Criteria pane does not appear in the MS Query window, choose **View**, **Query** from the MS Query Menu bar. In Figure 7-11, the criterion

AGE > '45'

has been selected. The results of the filter are immediately displayed in the bottom pane of the MS Query window.

The data view can be manipulated within MS Query, or the data can be copied to an Excel worksheet. Once the data become part of an Excel worksheet, they can be modified and the database functions described in this chapter can be used. To copy the

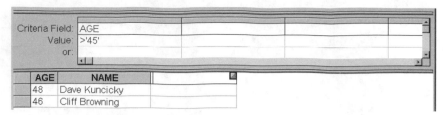

Figure 7.11.   Filtering a query

data to an Excel worksheet, choose **File**, **Return Data to Microsoft Excel** from the MS Query Menu bar.

Microsoft Query is a full-featured, stand-alone program and contains many features that are not explained here. For example, a *Structured Query Language (SQL)* statement may be created and used to query remote data. The criterion used in Figure 7-11 is equivalent to the following SQL statement:

```
SELECT staff.AGE, staff.NAME

FROM `C:\Dave`\staff.dbf staff

WHERE (staff.AGE>'45')
```

SUMMARY

In this chapter, the Excel database features were described. These include several search and filter tools, such as the AutoFilter and the Advanced Filter. Data may be retrieved from external databases by using the stand alone program Microsoft Query.

KEY TERMS

database	filtering
database management system	list
DBMS	record
field	sort key
field name	Structured Query Language (SQL)

Problems

1. Use the AutoFilter function to list only students whose GPA is less than 3.0.
2. Use the AutoFilter function to list all <u>Juniors</u> whose last name begins with S.
3. Use the Advanced Filter to list all <u>Seniors</u> with a GPA less than 2.9 and all <u>Juniors</u> with a GPA greater than 3.5. Can this list be created using the AutoFilter?
4. Create a new field titled <u>Full Name</u> that contains the full name of each student. Use the CONCATENATE function to accomplish this task. Be sure to include the space between the names, .e.g., <u>Frank Washington</u>, not <u>FrankWashington</u>.
5. Locate or create an external database using dBase, Access, Paradox, or FoxPro. Import the database into an Excel worksheet using Microsoft Query.

8

Collaborating with Other Engineers

8.1 THE COLLABORATIVE DESIGN PROCESS

This chapter focuses on workbook preparation in a team environment. Engineering design is the process of devising an effective, efficient solution to a problem. The solution may take the form of a component, a system, or a process. Engineers generally solve problems by collaborating with others as a member of a team. As a student, you will undoubtedly be asked to participate in collaborative projects with other students.

You may or may not have much experience working on a team. If a team works together effectively, more can be accomplished by the team than through any individual effort (or even the sum of all the individual efforts). If the team members do not work together effectively, however, the group can become mired in power struggles and dissension. When this occurs, one of two things usually happens: Either the team makes little progress towards its goals, or a small subgroup of the team takes charge and does all of the work.

Some guidelines for being an effective team member are presented at the end of this chapter in the "Professional Success."

8.1.1 Microsoft Excel and Collaboration

The ability to work well on a team can best be learned by participating on a successful team. Microsoft Excel includes several tools that can help to solve one of the most burdensome technical tasks of group collaboration: the preparation of the team document. In the past, collaborative preparation of documents has been extremely difficult. The result has been that the task is usually assigned to one or two team members. New features of Microsoft Excel make it feasible for the whole team to participate in the composition and

OBJECTIVES

After reading this chapter, you should be able to:

- Track revisions
- Share workbooks
- Insert comments
- Transfer worksheet data to and from other applications
- Use a password to restrict opening a file
- Use a password to restrict writing to a file
- Use a password to restrict access to a workbook
- Use a password to restrict access to a worksheet

APPLICATION

The best example of a modern, large collaborative engineering project is the design of the Boeing 777. This jetliner, which has been nicknamed the "triple seven," was designed to fly faster and farther with lower operating cost than any of its competitors.

Over 10,000 people worked on the $5 billion project. An international consortium of companies was brought together that included airplane developers, manufacturers, suppliers, and representatives of airline customers. Participants from the diverse disciplines, cultures, and corporations worked on more than 200 teams to design the 777 over a five-year period.

The teams that designed the Boeing 777 designed all parts of the airplane digitally and tested the parts to make sure they fit together without ever building a physical mockup of the airplane. Would you like to get an understanding of the magnitude of the problems that were encountered trying to get 10,000 participants from diverse disciplines, cultures, and corporations to share documents during the design of the 777? Start by reading the document titled *Some Groupware Challenges Experienced at Boeing by Steven E. Poltrock.*[1]

1. Poltrock. S. E. Some Groupware Challenges Experienced at Boeing, The Boeing Company, Seattle, WA. URL: http://orgwis.gmd.de/~prinz/cscw96ws/poltrock.html [web page, accessed 6/15/98].

revision of a document. Learning to use these features will require some time and practice on your team's part. The rewards will be well worth the effort.

8.2 TRACKING CHANGES

One problem that arises in team preparation of documents is keeping track of revisions. For example, one team member may be given the task of revising a section of the project. After the revisions are made, the team will meet and approve some or all of the revisions. Then one of the team members will incorporate the accepted changes into the document.

Excel has a feature called *Track Changes* that not only will mark revisions, but also will keep track of who is making each revision. The worksheet may be printed showing both the original text and the new revisions. The revisions then may be globally accepted or selectively accepted into the document.

To turn on the *Tracking Changes* feature, choose **Tools**, **Track Changes**, **Highlight Changes**. The Highlight Changes dialog box depicted in Figure 8-1 should appear on the screen.

Check the box labeled <u>Track changes while editing</u>. This will share the workbook and will turn on tracking. The next three boxes and drop-down lists allow you to limit the changes that are highlighted by time, user, and worksheet region. The bottom two items let you decide where to display the revisions. If you select the box labeled <u>Highlight changes on screen</u>, the revisions will be highlighted in the current worksheet, as depicted in Figure 8-2. The alternative selection saves the changes to a new worksheet.

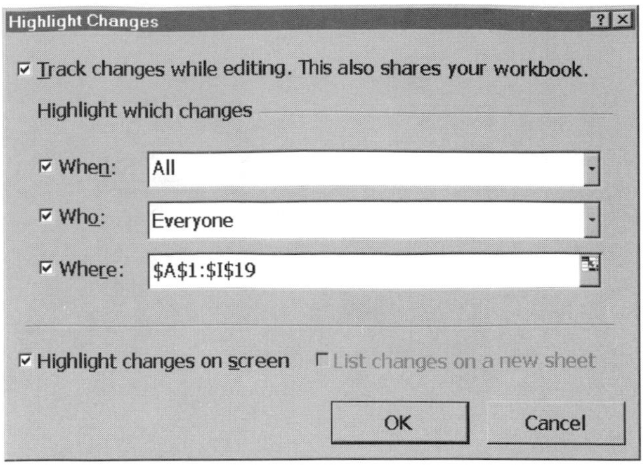

Figure 8.1. The Highlight Changes dialog box

Check the box labeled <u>Highlight changes on screen</u>. Now type a few words in your worksheet and notice what happens. Any modified cells are outlined in blue, and a small blue tab is placed in the upper left corner of the cell. These are called *revision marks*. (See Figure 8-2.)

	A	B	C
1	32	-2.5714	6.6122
2	14	-20.5714	423.1837
3	53	18.4286	339.6122
4	26	-8.5714	73.4694
5	18	-16.5714	274.6122
6	45	10.4286	108.7551
7	54	19.4286	377.4694

Figure 8.2. Examples of highlighted, revised cells

As you review a document for which other team members have made revisions, you can see who made the revision, the date and time of the revision, and the previous contents of the cell. The identity feature works only if each reviewer has given an identity to the Excel application. To identify yourself to Excel, choose **Tools**, **Options**, and then select the tab labeled **General**. Type your name in the box labeled **User Name**.

Your identity will be attached to any revisions that you make to a worksheet. You can test the feature by making a few revisions and then moving the mouse cursor over a section that has revision marks. A small box should appear that displays the reviewer's name along with a date-and-time stamp. An example is depicted in Figure 8-3.

8.2.1 Incorporating or Rejecting Revisions

Revision marks are not wholly incorporated into the document until they are reviewed and either accepted or rejected. Revision marks can be reviewed by using the Accept or Reject Changes dialog box. To accept or reject changes, choose **Tools**, **Track**

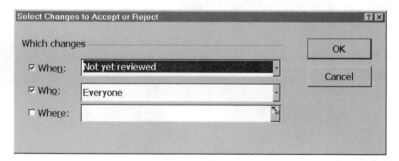

	A	B	C	D
1	32	-2.5714	6.6122	
2	14	-20.5714	423.1837	
3	53	18.4286	339.6122	
4	26	-8.5714	73.4694	
5	18	-16.5714	274.6122	
6	45	10.4286	108.7551	
7	54			
8				
9				

Professor Gooding, 6/13/98 2:40 PM:
Changed cell A7 from '32' to '54'.

Figure 8.3. Sample revision marks with reviewer's name

Changes, and **Accept or Reject Changes** from the Menu bar. The Accept or Reject Changes dialog box depicted in Figure 8-4 should appear. The three boxes and drop-down lists allow you to select which changes to review. Choose a time, reviewer, and region to review, and then choose **OK**. Excel will guide you through each selected revision and give you the opportunity to accept or reject the revision.

Figure 8.4. The Accept or Reject Changes dialog box

8.3 COMMENTS

At times, a reviewer may want to attach notes or *comments* to a cell without changing the contents of the cell. To add a comment to a cell, select the cell and choose **Insert**, **Comment** from the Menu bar. A Comment box depicted in Figure 8-5 will appear. Type in a comment. Then click outside of the Comment box to exit. A cell that is attached to a comment will be marked with a small red tab in the upper right-hand corner.

Comments may be reviewed, edited, or deleted by two methods. To review a single comment, select the cell holding the comment and click the right mouse button. The drop-down Quick Edit menu will display several new items, as depicted in Figure 8-6.

If there are a large number of comments to review, then open the Reviewing toolbar by choosing **View**, **Toolbars** from the Menu bar, and check the box labeled Reviewing.

The Reviewing toolbar contains buttons for moving through the comments in a worksheet one at a time, selectively viewing, editing, or deleting each comment. The Reviewing toolbar is displayed in Figure 8-7. The buttons related to comments are described in Table 8-1.

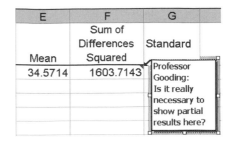

Figure 8.5. An example of a Comment box

Figure 8.6. Comment-related items on the Quick Edit menu

Figure 8.7. The Reviewing toolbar

TABLE 8-1. The Comment buttons on the Reviewing toolbar

BUTTON	ACTION
	Edit comment.
	Move to previous comment.
	Move to next comment.
	Show current comment.
	Show all comments.
	Delete current comment.

8.4 MAINTAINING SHARED WORKBOOKS

Excel provides a mechanism for several users to simultaneously share a workbook over a network. A shared workbook must reside in a shared folder on the network. Other access restrictions may apply, depending on your local network setup; see your network administrator for assistance with sharing folders. One other restriction to sharing workbooks is that all group members must be using Excel 97. Previous versions of Excel do not support this feature.

Once you are able to share a workbook, different users can view and modify the workbook at the same time. Sharing a document clearly requires some protocol among the group to keep several users from overwriting each other's work. Sharing a workbook is most effective if simultaneous users edit different sections of the workbook. Excel can be set up to keep a history of changes to a shared workbook, and previous versions may be recalled if necessary.

To share a workbook, choose **Tools**, **Share Workbook** from the Menu bar. The Share Workbook dialog box will appear. Select the **Editing** tab as depicted in Figure 8-8. By checking the box labeled <u>Allow changes…</u>, you permit the workbook to be shared. Once the workbook is shared you can utilize the **Editing** tab is useful to see who is currently using the workbook.

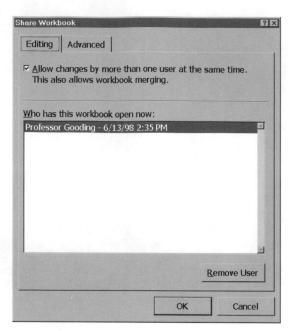

Figure 8.8. The Editing tab on the Share Workbook dialog box

8.4.1 Keeping a Change History

Excel can keep a log of changes made by each user who is a member of a shared work group. The log of changes is called a *change history*. Choose the **Advanced** tab on the Share Workbook dialog box. (See Figure 8-9.) Then choose the section labeled <u>Track changes</u>. The length of time to keep a change history can be selected in this section, or the change history can be turned off. One reason to turn off the change history or to keep its duration short is to limit the size of the workbook. A change history can significantly increase the disk space required to store a workbook. There is a trade-off between safety and storage requirements. Use of the change history feature is not a substitute for regularly backing up a workbook to some other medium, such as a floppy or tape.

Note: If a shared workbook has sharing turned off, then the change history is automatically deleted.

8.4.2 Timed Updates

The second section of the Shared Workbook dialog box's Advanced Setting tab is used to specify when changes are updated, so that other users may see them. The first selection specifies that your changes will be updated to the group whenever you save the file. Alternatively, you can choose to have the changes automatically update the other users' view of the workbook every few minutes.

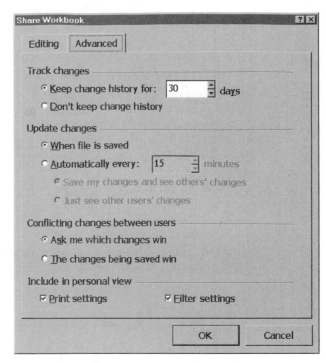

Figure 8.9. The Advanced tab on the Share Workbook dialog box

8.4.3 Managing Conflicts

If you are about to save a workbook, some of your changes may conflict with pending changes from another user. The third section of the Advanced Setting tab allows you to specify how you want to resolve conflicts, if at all. If you choose the first opeation, titled <u>Ask me which changes win</u>, then the Resolve Conflicts dialog box will appear when you save the file. You will be prompted to resolve each conflict. If you don't want to resolve conflicts when you save a shared workbook, then click the item titled <u>The changes being saved win</u>. The last user to save conflicting changes wins.

8.4.4 Personal Views

The last section of the Advanced Setting tab allows the creation of personal printer or filter settings. When the workbook is saved, a separate personal view is saved for each user.

8.4.5 Merging Workbooks

Group members do not always have access to the same network. One scenario that occurs when groups collaborate on a workbook is that each member takes a copy of the workbook. Each group member works separately on the workbook, and later the workbooks are merged into a single document.

Copies of a workbook can be revised and merged only if a change history is being maintained. Be sure to set a sufficient length of time for the change history so that the history doesn't expire before the workbook copies are merged. The number of days is set in the Share Workbook dialog box. (See Figure 8-9.)

To merge several copies of a workbook, open the first copy and then choose **Tools**, **Merge Workbooks** from the Menu bar. You will be prompted to choose a file to merge. Continue to merge files until all of the copies have been merged into one workbook. Follow the instructions in the next section to view the history of all changes that have been made.

8.4.6 Viewing the History of Changes

To view a change history, choose **Tools**, **Track Changes**, **Highlight Changes** from the Menu bar. The Highlight Changes dialog box depicted in Figure 8-1 will appear. Use the boxes and lists titled <u>Who</u>, <u>When</u>, and <u>Where</u> to limit the history list, or leave them unchecked to view all history entries.

If the box labeled <u>Highlight changes on screen</u> is checked, changes can be viewed one at a time by holding the mouse cursor over a changed cell. A pop-up box will appear with the change listed.

If the box titled <u>List changes on a new sheet</u> is checked, the history list will be displayed on a separate workshee,t as depicted in Figure 8-10. This feature is useful if multiple changes have been made to a cell.

Note in Figure 8-10 that cell A3 has been changed twice. Also note that the Auto-Filter feature has been turned on. This allows you to filter changes in the history list. The AutoFilter feature is described in Chapter 7.

	A	B	C	D	E	F	G	H	I
1	Action Number ▾	Date ▾	Time ▾	Who ▾	Change ▾	Sheet ▾	Range ▾	New Value ▾	Old Value ▾
2	1	6/13/98	2:42 PM	Professor Gooding	Cell Change	Sheet1	A3	53	52
3	2	6/13/98	2:42 PM	Professor Gooding	Cell Change	Sheet1	A7	54	32
4	3	6/13/98	3:10 PM	Dr. Doolittle	Cell Change	Sheet1	A3	43	53

Figure 8.10.   Reviewing a change history

8.4.7 Restrictions for Shared Workbooks

Some features of Excel cannot be used while a workbook is being shared. All features can be used if the workbook has sharing turned off. The disadvantage of turning off sharing is that the change history is deleted. The following are some of the features of Excel that cannot be used when sharing is in effect:

- Creation, modification, or deletion of passwords. Passwords should be set up before the workbook is shared.
- Deletion of worksheets.
- Insertion or modification of charts, pictures, or hyperlinks.
- Insertion or deletion of regions of cells. Single rows or columns can be deleted.
- Creation of data tables or pivot tables.
- Insertion of automatic subtotals.

Other restrictions for shared workbooks, can be viewed by choosing **Help**, **Contents and Index** from the Menu bar. Select the **Index** tab from the Help dialog box. Type the keywords <u>shared workbook</u>, select **limitations** from the resulting list of keywords, and choose **Display**. A help box will appear that describes all of the limitations that apply during the sharing of a workbook.

8.5 PASSWORD PROTECTION

Several levels of protection exist for workbooks that reside on a network. Your personal file space may be protected by the network operating system. The folder in which the workbook resides may be protected. The methods that are discussed next apply only to a single workbook. These methods, in and of themselves, will not prevent another user from copying your workbook. You should discuss general file protection issues with your local system or network administrator.

One way to limit access to a shared workbook is with password protection. A variety of types of password will be discussed. In each case, be sure to write down your password. If you lose a password, you will not be able to retrieve your work.

8.5.1 Open Protection

A password can be set that restricts a user from opening a file. This means that an unauthorized user cannot read or print the file by using Excel. This type of access is called *open access,* since it protects a file from being opened. A user may still be able to copy the file and interpret it using some other program, however. To set a password for open access, choose **File**, **Save As**, **Options** from the Menu bar. The Save Options dialog box depicted in Figure 8-11 will appear.

The first option, titled <u>Always create backup</u>, specifies that Excel should create a backup copy of your workbook every time it is saved. Unless you are extremely short of disk space, this is an excellent option!

To restrict open access, type a password in the box titled <u>Password to open</u>. You will be prompted to type the password a second time for verification. Note that Excel uses case-sensitive passwords.

Hint: One of the most common reasons that a password seems to suddenly stop working is that you have the caps lock key turned on.

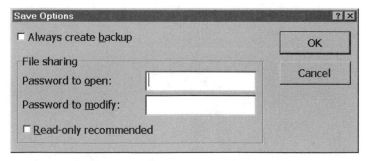

Figure 8.11. The Save Options dialog box

8.5.2 Write Protection

There may be times when you want to allow read access to other people but you do not want anyone to be able to modify your original file. This type of protection is called *write access*. To set a write access password, open the Save Options dialog box depicted in Figure 8-11. Type a password in the box titled <u>Password to modify</u>. You will be prompted to type the password a second time for verification.

The next time you attempt to open the file, the Password dialog box depicted in Figure 8-12 will appear. You will be prompted for a password if you want to open the file for write access. A password is not needed to open the file for reading only. Note that a user can open a write-protected file as a read-only file, then save it under a different name. The new file can be modified by the user without a password.

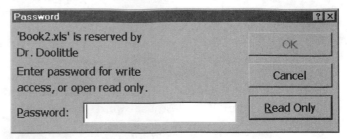

Figure 8.12. The Password dialog box

8.5.3 Sheet Protection

Protection can be finely tuned. This feature is convenient when you are sharing a document, since a user can protect part of the workbook, but leave some sections available to others. Another option is to protect only the structure or window configuration of a workbook, but allow others to modify the cell contents. Even if a document is not being shared, you may want to protect it. Once you have completed part of a worksheet, you may want to protect it merely to prevent yourself from inadvertently modifying that section.

To protect a single worksheet within a workbook, choose **Tools**, **Protection**, **Protect Sheet** from the Menu bar. The Protect Sheet dialog box depicted in Figure 8-13 will appear. From this box, you can choose to protect the contents of cells, objects such as drawings, charts, or scenario definitions.

You may want to keep certain regions unprotected in a protected worksheet. To keep a selected region unprotected, before protecting the worksheet, select the region and choose **Format**, **Cells**. Select the **Protection** tab, and then remove the check from the item labeled <u>Locked</u>. Now, after the worksheet is protected, the unlocked region will be exempted and can still be modified.

Figure 8.13. The Protect Sheet dialog box

8.5.4 Workbook Protection

An entire workbook can be protected in a manner similar to protecting a worksheet. Choose **Tools**, **Protection**, **Protect Workbook** from the Menu bar. The Protect Workbook dialog box depicted in Figure 8-14 will appear. From the Protect Workbook dialog box, you can choose to protect the workbook structure or the window setup. Protecting the workbook structure means that worksheets cannot be added, deleted, moved, or hidden. Protecting the window setup means that that the workbook windows cannot be deleted, resized, or moved.

Figure 8.14.   The Protect Workbook dialog box

8.6 IMPORTING AND EXPORTING DATA

Several methods for importing data are discussed in other chapters in this book. In Chapter 7, the Microsoft Query program is introduced. Microsoft Query enables the selective retrieval of data from external database files such as Oracle, dBase, or Paradox. In Chapter 9, Web queries are discussed. A Web query is designed to retrieve data from sites on the World Wide Web, using your default browser as an interface.

A number of types of file formats may be imported directly by choosing **File**, **Open** from the Menu bar. Click on the arrow to the right of the box labeled <u>Files of type</u>. The small drop-down menu depicted in Figure 8-15 will appear.

An external file may be opened and viewed within Excel. If the file is modified, Excel will ask if you want to save it in its original format or in Excel format. If the file is saved in Excel format, the external program (e.g., dBase) will not be able to view the changes. If the file is saved in the external format (e.g., dBase), some Excel formatting may be lost. For example, formulas and macros may not be translated into the external format.

What do these announcements have in common? Teamwork! The ability of an engineer to work well in the team environment has as much to do with professional success as the engineer's scientific and technical skills. Few engineering accomplishments are produced in isolation. The following guidelines may help as you begin to conduct and participate in team meetings.

Decision Making.   Attempt to make decisions by consensus. If that fails due to a single member who disagrees, move to consensus minus one.

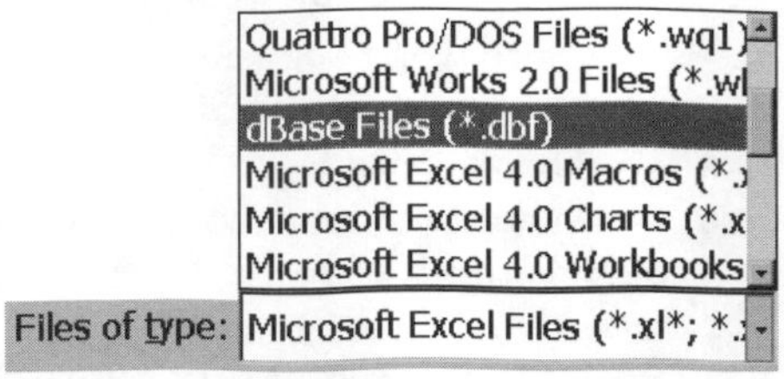

Figure 8.15. Opening external types of files

PROFESSIONAL SUCCESS:
GUIDELINES FOR TEAM MEETINGS

Consider the Following announcements for job positions:

Mechanical Engineer: We have an immediate need for an engineer to interface an engineering automation & optimization environment with a variety of CAD and CAE systems.... B.S./M.S. in mechanical engineering .... Good communication skills and a strong interest to *interact with customers in problem-solving situations* is a must.

Electrical Engineer: Required degree: BSEE+. Perform audio subsystem validation to verify prototypes throughout the product development program cycle. *Must work well in a team environment.*

Aerospace Engineer: Applicants selected may be subject to a government security investigation and must meet eligibility requirements for access to classified information. You *must be a team player* and possess excellent written and oral communication skills.

Industrial Engineer: Investigate manufacturing processes in continuous improvement environment; recommend refinements. Design process equipment to improve processes. Must be highly skilled at planning/managing and be *able to sell ideas to team members* and company management.

Extrusion Engineer: This manufacturer of fiber optics is seeking an extrusion engineer who can handle the majority of the technical issues in production. The successful candidate must be able to *work with other disciplines in a team atmosphere* of mutual support.

Software Engineer: Looking for a well-rounded software engineer with strong experience in object-oriented design and GUI development... Must be a highly motivated self-starter *who works well in a team environment.*

Confidentiality. Respect the confidentiality of other members. Lay ground rules about what material, if any, is to be treated confidentially. In the world of business and government contracts, you may be asked to sign a *confidentiality agreement*. This legal document specifies what materials are protected.

Attention. Listen Actively. Ask questions or request clarification of other members' comments. Summarize of important points that other team members have made. Acknowledge the fact that you have understood. Try not to mentally rehearse what you are going to say when others are speaking.

Preparation. Be adequately prepared for the meeting.

Punctuality. Be on time. If you are 10 minutes late and there are six other members in the group, then you are wasting one person hour of time!

Ensure Active Contribution.   If not all team members are contributing and actively participating, then something is wrong with the group process. Stop the meeting, and take time to get everyone involved before proceeding.

Record Keeping.   Someone on the team should be keeping records of team meetings.

Flexibility.   One of the aspects of working on a team is that you win some and you lose some. Not every one of your ideas will be accepted by the group. Be prepared to think of creative solutions that every team member can accept.

Dynamics.   Help improve relationships among the team members. Do not dominate the meeting or let another member dominate the meeting. If this cannot be resolved within the group, enlist the help of an outside *facilitator*. A facilitator is a non—group member who does not deal with the content of the group issues. Rather , the facilitator helps smooth the group *process*. One of the roles of a facilitator is to prevent any single team member from dominating the others.

Quorum.   Establish at the onset what a team quorum will be. Do not hold team meetings unless a quorum is present.

SUMMARY

In this chapter, the tools that Excel furnishes for collaboration were discussed. These include methods for tracking revisions, sharing workbooks, inserting comments, and importing data from other applications. The chapter also explained the various methods of data protection that Excel provides.

KEY TERMS

change history	open accessc
comments	revision marks
confidentiality agreement	tracking changes
facilitator	write access

Problems

1. Create a workbook and make several copies of it. Make revisions to each of the separate documents using the methods described in Section 8.4. Merge the revised documents into a single document by opening the copy of the shared workbook into which you want to merge changes from another workbook file on disk. Choose **Tools**, **Merge Workbooks**. Select the shared workbook to be merged, and click OK. Repeat these steps for each copy that is to be merged. You will be guided through the process of accepting and rejecting revisions.

2. Turn on the AutoSave feature to automatically save your document every two minutes. Is a new version created every two minutes? You can check this by closing Excel without manually saving your changes and testing to see whether the changes have been saved.

3. Do the password protection mechanisms discussed in this chapter prevent another student from making a copy of your paper? Do any of the protection methods presented in this chapter prevent someone from printing your document without knowing the password? If so, which ones?

4. Turn sharing on and create a change history for a workbook. Then turn sharing off and see if the change history is actually deleted.

5. Set the change history timer in the Share Workbook dialog box (see Figure 8-9) to one day. Wait more than 24 hours and see if the history really expires.

9

Excel and the World Wide Web

9.1 ENGINEERING AND THE INTERNET

The Internet is one of the primary means of communication for scientists and engineers. Correspondence through electronic mail, the transfer of data and software via electronic file transfer, and research using on-line search engines and databases are everyday occurrences for engineers. The World Wide Web (WWW, or simply, the Web) is a collection of technologies for publishing, sending, and obtaining information over the Internet. There are two new essential skills to be learned by every engineering student. First, every student must gain fluency in searching, locating, and retrieving relevant technical information from the WWW. Second, every engineering student must learn how to post written documents to the WWW. The ability to present technical results via the WWW is an essential communication skill for today's engineer.

9.2 ACCESSING THE WORLD WIDE WEB FROM WITHIN EXCEL

To access the Internet from within Excel, your computer must be connected to the Internet. If you are in a computer lab at school, then the computer may be connected to a local area network (LAN) through a network card. The LAN may or may not be connected to the Internet. Ask your lab manager or system administrator for details on your connection. If your computer is not directly connected to a LAN, you can still access the Internet using a modem. This is called *dial-up networking*. For information about dial-up networking, access the Help section on the Task bar (if you are using Microsoft Windows 95 or NT). Locate the topic titled Dial-

OBJECTIVES

After reading this chapter, you should be able to:

- Access the World Wide Web from within an Excel worksheet
- Obtain extra templates and other add-ins from the WWW
- Retrieve files from FTP and HTTP servers into a local worksheet
- Use the Web Query feature to import Excel data from the WWW
- Create hyperlinks in a worksheet
- Convert of Excel documents to HTML

The World Wide Web holds a wealth of information about your new profession. Take some time to visit the professional societies that represent your discipline. The following URLs represent a few of the national and international organizations that are on-line.

Accreditation Board for Engineering and Technology (ABET)	http://www.abet.org/ABET.html
American Institute for Aeronautics and Astronautics (AIAA)	http://www.aiaa.org
American Institute of Chemical Engineers (AICHE)	http://www.aiche.org
American Society of Civil Engineers (ASCE)	http://www.asce.org
American Society of Engineering Education (ASEE)	http://www.asee.org/asee
American Society of Mechanical Engineers (ASME)	http://www.asme.org
American Society of Naval Engineers(ASNE)	http://www.jhuapl.edu/ASNE
Institute of Electrical and Electronic Engineering (IEEE)	http://www.ieee.org
National Society of Black Engineers (NSBE)	http://www.nsbe.org
National Society of Professional Engineers (NSPE)	http://www.nspe.org
Society of Women Engineers (SWE)	http://www.swe.org

Up Networking. The computer systems group at your college may be able to help you with some of the details, such as the assignment of an IP address, netmasks, etc. During the rest of this chapter, it is assumed that your computer is connected to the Internet.

Access the Web toolbar by choosing the 🌐 icon on the Standard toolbar. The Web toolbar depicted in Figure 9-1 should appear. To open a Web page or local Web document, choose Go ▾ from the Web toolbar and select the open button 📂 from the drop down menu. The Open Internet Address dialog box depicted in Figure 9-2 will appear.

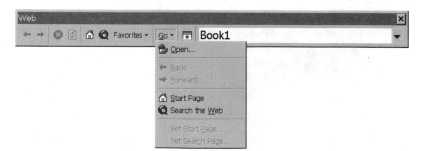

Figure 9.1.   The Web toolbar

From the Open Internet Address dialog box, you can type or choose a remote or local Web page. The address of a remote Web page is called a *Uniform Resource Locator,* or *URL.* A URL takes the form

http://www.eng.fsu.edu/succeed/succeed.html

where

http	stands for HyperText Transfer Protocol,
www.eng.fsu.edu	is the name of a Web server, and
succeed/succeed.html	is the path name of a Web page on that server

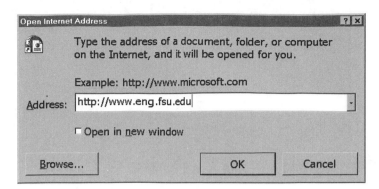

Figure 9.2. The Open Internet Address dialog box

Figure 9-2 shows the URL for the Florida A&M University–Florida State University (FAMU–FSU) College of Engineering Web page. You can also type in the path and name of a local Web document, or select the local document by using the arrow to the right of the box labeled **Address**.

A Web document is written in a markup language called *HyperText Mark-up Language,* or *HTML.* The HTML document is usually viewed using an application called a Web browser. The browser interprets the HTML document and presents it as text, graphics, animations, sounds, etc.

Excel acts as a front end to a Web browser. When you click **OK**, the selected page will be displayed through your default Web browser, such as Microsoft's Internet Explorer or Netscape Navigator. Since you may already be familiar with your favorite browser, you may want to use that browser's features directly, instead of the Excel front end, for general Web browsing.

PRACTICE!

Practice accessing several Web sites and files from within Excel, and familiarize yourself with the Web toolbar. Choose the arrow on the right-hand side of the toolbar, and the pop-up scroll box depicted in Figure 9-3 will appear. This box keeps a history of URLs that you have previously accessed. The horizontal arrows ← ⇒ page backward and forward through sites visited in the current session. The **Favorites** item is a bookmark feature. The Web Toolbar button 🔲 moves the Web Toolbar from the top of the screen to a separate box (and vice versa).

```
http://www.eng.fsu.edu
http://www.eng.fsu.edu/net98
C:\Dave\Fsu\Prentice-Hall\Excel\Stude
http://www.microsoft.com
http://www.cs.fsu.edu/~jtbauer/cis54(
D:\net98\lectures\program\18 Rpcgen
http://www.hp.com/pso/frames/servic
http://iq.orst.edu/sysadm/
http://iq.orst.edusy/sysadm/
http://www.eng.fsu.edu/~kuncick/uni
http://cq-pan.cqu.edu.au/david-jones
http://science.cqu.edu.au/mc/Academ
```

Web
← ⇒ ⊗ 🔲 🏠 🔍 Favorites ▾ Go ▾ 🔲 \Prentice-Hall\Excel\Mean Median.xls ▾

Figure 9.3. Viewing previously accessed sites

9.3 WEB SITES RELATED TO MICROSOFT EXCEL

Microsoft maintains a Web site with sections that are specifically applicable to Excel users. The primary page for Microsoft is located at

http://www.microsoft.com

Within the Microsoft site is an Excel 97 tutorial, information on the company's products, and a number of free add-ins, patches, and templates.

9.4 RETRIEVING DATA FROM A WEB PAGE

The Web Query feature retrieves data from an external source over the Web and places the data in a local Excel worksheet. Before proceeding, open a new worksheet. Then select **Choose Data**, **Get External Data**, **Run Web Query** from the Menu bar. The Run Query dialog box depicted in Figure 9-4 will appear. Several sample queries are provided with the standard Excel 97 installation. If you would like to download more Web queries, choose the item titled **Get More Web Queries**.

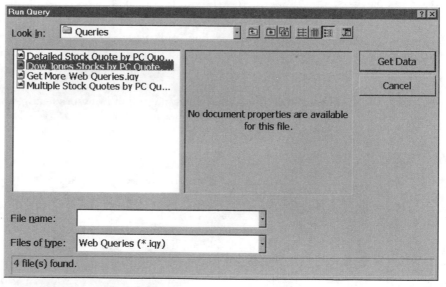

Figure 9.4. The Run Query dialog box

A *Web query* is a formatted text file. The contents of the Dow Jones query are displayed in Figure 9-5. The effect of executing the query is to access the Web server at the URL http://webservices.pcquote.com and execute the CGI program named excel-dow.exe. The results are returned and displayed in your local Excel worksheet.

Choose the item **Dow Jones Stock Quotes by PC Query, Inc.** The Returning External Data dialog box depicted in Figure 9-6 will appear. From this dialog box, you can choose to place the returned data in the current worksheet, a new worksheet, or a pivot table report. Choosing the **Properties** button will allow you to modify a number of query options, such as the ability to automatically refresh or update the data.

```
WEB

1

http://webservices.pcquote.com/cgi-bin/exceldow.exe?
```

Figure 9.5. Contents of Dow Jones Stock Query by PC Query, Inc.

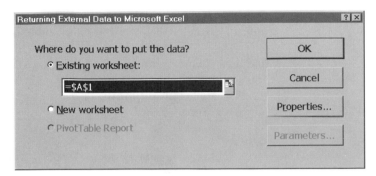

Figure 9.6. The Returning External Data dialog box

If the query supports input parameters, then the **Parameters** button will be active. An example of a parameter would be the addition of a particular stock symbol for a stock that you wish to download.

Part of the results from the previous query are displayed in Figure 9-7. Note that the AutoFilter feature has been automatically turned on. This allows you to use the filtering and searching techniques that you learned in Chapter 7. In Figure 9-7, column D has been selected to filter the top 10 items for the field titled Net Change.

	A	B	C	D	E	F
	Company Name & Symbol	Category	Last Price	Net Change	Open	High
1						
7	CHEVRON CORP(CHV)	INDUSTRIALS	81 5/16	1 3/16	80 5/8	81 7/16
9	Name Not Available(DIS)	INDUSTRIALS	112 7/8	3 5/16	112 15/16	113 3/8
10	Name Not Available(EK)	INDUSTRIALS	69 5/16	5/16	69 3/8	69 3/8
15	IBM INTERNATIONAL BUSINESS MACHINES(IBM)	INDUSTRIALS	116 3/4	11/16	116 3/4	117 7/16
22	PHILIP MORRIS COMPANIES INC(MO)	INDUSTRIALS	36 5/8	3/8	36 15/16	37 1/16

Figure 9.7. Results from PC Query, Inc., Dow Quote

9.5 ACCESSING FTP SITES FROM WITHIN EXCEL

A workbook can be opened from a remote site using *FTP*, or *File Transfer Protocol*. Before FTP can be used, you must be connected to the Internet using a dial-up connection (via a modem) or a direct network connection through a (network card). In addition, you must add the FTP site to a list of Internet sites.

To add an FTP site to your list, choose **File**, **Open**, and click the arrow on the right side of the box labeled **Look In** Choose **Internet Locations (FTP)** from the drop-down list. The Add/Modify FTP Locations dialog box depicted in Figure 9-8 will appear. From the dialog box, you can add, modify, and delete FTP locations.

If you have an account on the remote FTP site, then check the box labeled **User**, and type in your user name and password to the FTP site. If you do not have an account, you still may be able to login, because many sites accept anonymous FTP logins. Check the box labeled **Anonymous**, or use a username of anonymous and enter your e-mail address as the password.

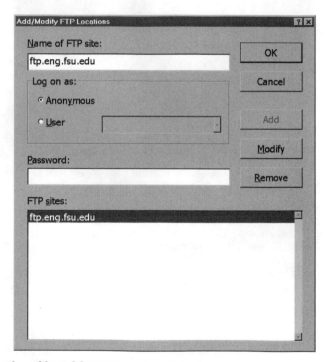

Figure 9.8.   The Add/Modify FTP Locations dialog box

PRACTICE!

Practice opening a remote workbook by typing adding the site ftp.eng.fsu.edu to your list as shown in Figure 9-8. Check the **Anonymous** login box. Open the site, and once you are anonymously logged in, select the **pub** folder, the **kuncick,** folder and then the **excel** folder. You should see a list of files similar to the list in Figure 9-9. Many of the worksheets used as examples in this book are stored at the FSU site. Download the worksheets, and use them to complete the examples in the book.

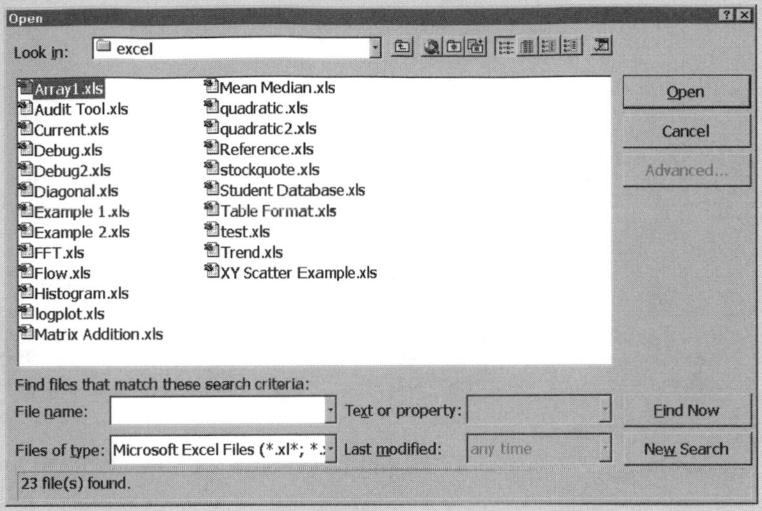

Figure 9.9. Opening a Remote Workbook using FTP

9.6 CREATING HYPERLINKS WITHIN A WORKSHEET

A *hyperlink* or simply, *link* can be thought of as a pointer to another document. When you click on a hyperlink, that document is immediately displayed. The linked document may be another Excel worksheet on your local computer, or it may be a document from another application, such as Microsoft Word. If the linked document belongs to anther application, then that application is started automatically for you.

A link may also point to a remote document that is retrieved from the World Wide Web using the HTTP or FTP protocol. As more computers are connected to the Internet and network speeds increase, the differences between accessing a local document and a remote document will diminish.

In this section, you will be shown how to create hyperlinks to several types of documents. The method for creating hyperlinks is the same for local or remote documents. The only difference is the address or path name of the document.

In the first example (see Figure 9-10), a link will be created from a worksheet containing grades to a student database. In this case, both documents are local Excel files. To link another worksheet or part of a worksheet to a cell, perform the following steps:

1. Select the cell that will contain the link (Student Database in the example in Figure 9-10).
2. Choose **Insert**, **Hyperlink** from the Menu bar, or choose the 🌐 button from the Standard toolbar.

	A	B	C	D	E	F
2	Clinton	Willie	89.3		Student Database	
3	Smith	Randolph	58.2			
4	Simpson	Susie	97.0		Mean=	81.1
5	Smith	Christine	77.4		Median=	83.35
6	Washingto	Frank	65.4			
7	Granger	Linda	99.4			

Figure 9.10. Example of a Hyperlink

3. The Edit Hyperlink dialog box depicted in Figure 9-11 will appear.
4. Type a local file path name, or use the **Browse** button to select a file.
5. You may limit the link to a name (i.e., a location) within the file by filling in the box labeled **Name Location In File**
6. Choose **OK**.

The contents of the selected cell will change color and will be underlined, indicating that the cell contains a hyperlink. Click the hyperlink and the referenced file will appear. You can toggle back and forth among the source and destination files by using the ← ⇒ arrows on the Web toolbar.

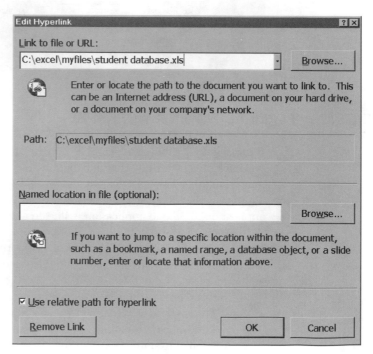

Figure 9.11. The Edit Hyperlink dialog box

A hyperlink may contain references to an HTTP or FTP server as well as to local files. An example of an HTTP URL is

```
http://www.eng.fsu.edu/net98
```

An example of an FTP URL is

`ftp://ftp.eng.fsu.edu/pub/kuncick/excel/Histogram.xls`

PRACTICE!

The FTP reference

`ftp://ftp.eng.fsu.edu/pub/kuncick/excel/Mean Median.xls`

will retrieve the data shown in Figure 9-10. The hyperlink will already be embedded in the worksheet, so you can experiment both with retrieving FTP files and with following hyperlinks.

After you have retrieved and opened the file <u>Mean Median.xls</u>, hold the mouse over the cell with the text <u>Student Database</u> (but don't press the mouse button). Note that the URL of the hyperlink appears in a small drop-down box, as depicted in Figure 9-12.

	E	F	G	H
	Student Database			
		ftp://ftp.eng.fsu.edu/pub/kuncick/excel/Student Database.xls		
	Mean=	81.1		
	Median=	83.35		

Figure 9.12. Viewing the URL of a hyperlink

Now click the right mouse button, and choose **Hyperlink**, **Edit Hyperlink** from the drop-down menu. The Edit Hyperlink dialog box will appear, and you can edit the hyperlink address. Try replacing the text <u>Student Database</u> with <u>Histogram</u>.

9.7 CONVERTING A WORKSHEET TO A WEB PAGE

Excel provides a Wizard to assist with the conversion of worksheets to HTML format.

To convert a worksheet to HTML, first open the worksheet that you want to convert, and then choose **File**, **Convert to HTML** from the Menu bar. The Internet Assistant Wizard (Step 1) will appear. One or more regions from the workbook will appear on the list in this dialog box. Add and remove selections until you are satisfied with the list. Then press **Next**.

Step 2 of the Internet Assistant Wizard will appear. You choices are to create a new HTML document or to insert the current selections into an existing HTML document. Choose **Create a New Document** and select the **Next** button.

Step 3 of the Internet Assistant Wizard, depicted in Figure 9-13 will appear. Type in a title, heading, and text for the Web page. You may also insert horizontal lines and add author information to the Web page in this dialog box. The example in Figure 9-13 is taken from the river flow data used in Chapter 5.

After you are finished with Step 3, select the **Next** button, and Step 4 of the Internet Assistant Wizard will appear. Type the path name pointing to where the HTML file will be saved, and choose **Finish**. Try viewing the finished page with your favorite Web browser.

Figure 9-14 shows the river flow data using the Netscape navigator browser. You can edit the HTML page with Microsoft Word 97 or any other text editor.

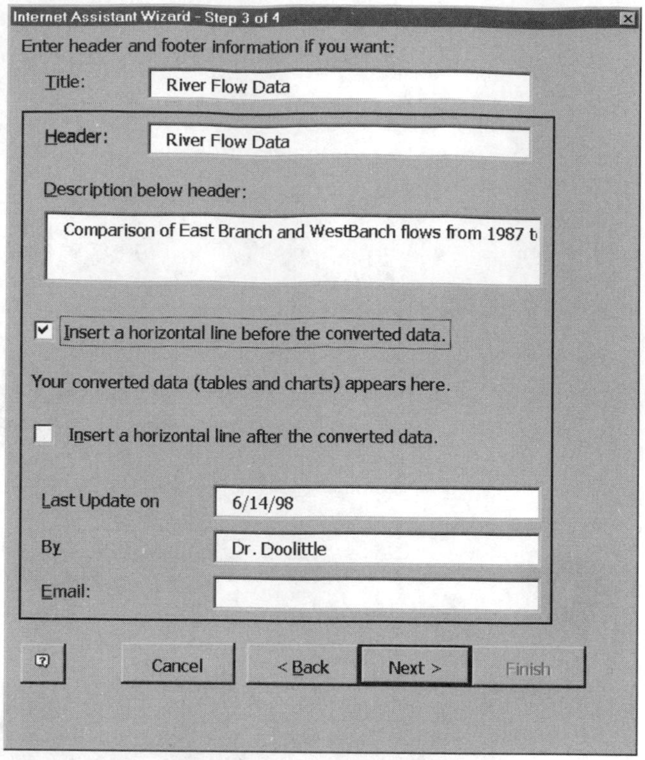

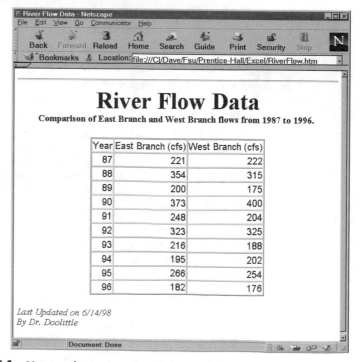

Figure 9.13. Step 3 of the Internet Assistant Wizard

Figure 9.14. Viewing the converted worksheet with Netscape

SUMMARY

This chapter examined the various ways that the World Wide Web and Excel can interface. The Web can be accessed from within an Excel worksheet. Excel files from remote HTTP and FTP files can be directly imported into a local worksheet. The Web Query feature automates the retrieval of remote data. Hyperlinks can be added to a worksheet, and Excel will convert worksheets to HTML using the Internet Assistant Wizard.

KEY TERMS

dial-up networking
File Transfer Protocol
FTP
HTML
hyperlink

HyperText Mark-up Language
link
Uniform Resource Locator
URL
Web Query

Problems

1. Use the help feature to read about the HYPERLINK function. Create a valid hyperlink in a worksheet using the HYPERLINK function.

2. One advantage of using the HYPERLINK function is that the link can depend on a conditional expression. Create an IF expression that links to www.netscape.com if cell A1 = Netscape and links to www.microsoft.com if cell A1 = Explorer.

Appendix A: Commonly Used Functions

ABS(*n*)	Returns the absolute value of a number
AND(*a*, *b*, ...)	Returns the logical AND of the arguments (TRUE if all arguments are TRUE, otherwise FALSE)
ASIN(*n*)	Returns the arcsine of *n* in radians
AVEDEV(*n1*, *n2*, ...)	Returns the average of the absolute deviations of the arguments from their mean
AVERAGE(*n1*, *n2*, ...)	Returns the arithmetic mean of its arguments
BIN2DEC(*n*)	Converts a binary number to decimal
BIN2HEX(*n*)	Converts a binary number to hexadecimal
BIN2OCT(*n*)	Converts a binary number to octal
CALL(...)	Calls a procedure in a DLL or code resource
CEILING(*n*, *sig*)	Rounds a number *n* up to the nearest integer (or nearest multiple of significance *sig*)
CHAR(*n*)	Returns the character represented by the number *n* in the computer's character set
CHIDIST(*x*, *df*)	Returns the one-tailed probability of the chi-squared distribution using *df* degrees of freedom
CLEAN(*text*)	Removes all nonprintable characters from *text*
COLUMN(*ref*)	Returns the column number of a reference
COLUMNS(*ref*)	Returns the number of columns in a reference
COMBIN(*n*, *r*)	Returns the number of combinations of *n* items choosing *r* items
COMPLEX(*real*, *imag*, *suffix*)	Converts real and imaginary coefficients into a complex number
CONCATENATE(*str1*, *str2*,...)	Concatenates the string arguments
CORREL(**A1**, **A2**)	Returns the correlation coefficients between two data sets
COS(*n*)	Returns the cosine of an angle
COUNTBLANK(*range*)	Counts the number of empty cells in a specified range
DEC2BIN(*n*, *p*)	Converts the decimal number *n* to binary using *p* places (or characters)

DELTA($n1, n2$)	Tests whether two number are equal
ISERROR(v)	Returns TRUE if value v is an error
ISNUMBER(v)	Returns TRUE if value v is a number
FACT(n)	Returns the factorial of n
FORECAST($x, known\ x's,$ $known\ y's$)	Predicts a future value along a linear trend
LN(n)	Returns the natural logarithm of n
MDETERM(**A**)	Returns the matrix determinant of array **A**
MEDIAN($n1, n2, ...$)	Returns the median of it arguments
MOD(n, d)	Returns the remainder after n is divided by d
OR($a, b, ...$)	Returns the logical OR of its arguments (TRUE if any argument is TRUE, FALSE if all arguments are FALSE)
PI()	Returns the value of pi to 15 digits of accuracy
POWER(n, p)	Returns the value of n raised to the power of p
PRODUCT($n1, n2, ...$)	Returns the product of its arguments
QUOTIENT(n,d)	Returns the integer portion of n divided by d
RADIANS(d)	Converts degrees to radians
RAND()	Returns an evenly distributed pseudo-random number >= 0 and < 1
ROUND(n, d)	Rounds n to d digits
ROW(ref)	Returns the row number of a reference
SIGN(n)	Returns the sign of a number n
SQRT(n)	Returns the square root of a number n
STDEVP($n1, n2, ...$)	Calculates the standard deviation of its arguments
SUM($n1, n2, ...$)	Returns the sum of its arguments
SUMSQ($n1, n2, ...$)	Returns the sum of the squares of its arguments
TAN(n)	Returns the tangent of an angle
TRANSPOSE(A)	Returns the transpose of an array
TREND($known\ y's,$ $known\ x's,\ new\ x's,$ $constant$)	Returns values along a linear trend by fitting a straight line using least squares
VARP($n1, n2, ...$)	Calculates the variance of its arguments

Index